SKYDIVING MADE FUN AND EASY

Tom Buchanan

McGraw-Hill

New York Chicago San Francisco Lisbon London Madrid
Mexico City Milan New Delhi San Juan Seoul
Singapore Sydney Toronto

The **McGraw·Hill** Companies

Library of Congress Cataloging-in-Publication Data

Buchanan, Tom (Tom E.)
   Jump! : skydiving made fun and easy / Tom Buchanan.
      p. cm.
     Includes index.
     ISBN 0-07-141068-6
     1. Skydiving. I. title.

   GV770 .B83 2003
   797.5'6—dc21

                                    2002038072

2 3 4 5 6 7 8 9 0 DOC/DOC 0 9 8 7 6 5 4 3

ISBN 0-07-141068-6

The sponsoring editor for this book was Shelley Ingram Carr, the editing supervisor was David E. Fogarty, and the production supervisor was Pamela A. Pelton. It was set in Goudy by Joanne Morbit and Paul Scozzari of McGraw-Hill Professional's composition unit, Hightstown, N.J.

Printed and bound by RR Donnelley.

This book is printed on recycled, acid-free paper containing a minimum of 50% recycled de-inked paper.

McGraw-Hill books are available at special quantity discounts to use as premiums and sales promotions, or for use in corporate training programs. For more information, please write to the Director of Special Sales, Professional Publishing, McGraw-Hill, Two Penn Plaza, New York, NY 10121-2298. Or contact your local bookstore.

Contents

About the Author   vii

Introduction   ix

Using This Book   vii
My First Jump   x
Contemporary Skydiving   xi

Chapter 1. Overview   1

What Kind of People Jump out of Perfectly Good
  Airplanes   3
Getting Started   5
Costs   7
Weather Issues   8
Liability and Insurance Concerns   11
Weight Limits   12
Medical Issues   14
Sensory Overload   18
Need for Speed   20
Emergencies   23

Chapter 2. Training Methods   25

Tandem Jumps   25
Static Line Jumps   30
Accelerated Free Fall   35
Making More Jumps   38
Video Training   39
Freefall Simulators   39

Parachute Simulators   44
Hybrid Training Programs   46

Chapter 3. Regulation and Instructor Certification   51

Federal Aviation Administration   51
United States Parachute Association   54
Skydive University   60
State Regulation   61
Tandem Manufacturers   62

Chapter 4. Psychology   65

Your Stress Level   65
A Model of Fear Management   66
Using Friends to Control Stress   75
Relaxing   76
Using Call-Outs   78
Converting Stress   79
Improving Your Life   80

Chapter 5. Equipment   83

A Complete Rig   83
Tandem Equipment   84
Student Parachute Rigs   90
Expert Skydiver Rigs   94
Automatic Activation Devices   96
Jumpsuits   99
Helmets   100
Goggles   101
Altimeters and Instruments   102
Log Books   105

Chapter 6. Aircraft   107

Piston versus Turbine   107
Common Jump Planes   110
Other Airplanes   117

Chapter 7. Understanding Risk **119**

USPA Reports   120
Tandem   126
Injuries   128
Other Activities   128
Relative Risk   130

Chapter 8. How Skydivers Fly **133**

Arch   134
Flying the Parachute   137
Formation Skydiving   137
Freefly   140
Speed Skydiving   142
Wingsuit Flight   144
Sky Surfing   146
High-Altitude Jumping   147
Parachute Swooping   148
CRW   150
Exhibition Jumping   152
BASE Jumping   152
Mixing It Up   154

Chapter 9. All about Drop Zones **157**

Organization   157
Spectators   159
Social Activity   160
Large Drop Zones   162
Small Drop Zones   165
Making It All Work   168

Chapter 10. Making Your Decision **177**

Getting Started   179
Checking the Web   181
Using the Phone   182
Comparing the Answers   191
Making the Decision   192

**Appendix. Frequently Asked Questions
and Answers   195**

Notes   205

Glossary   208

Acknowledgments   221

Index   223

Tom Buchanan holds Instructor ratings issued by The United States Parachute Association in Accelerated Freefall, static line, instructor assisted freefall, and tandem training. He has earned tandem equipment ratings covering the Vector, Sigma, Racer, and Dual Hawk systems. Mr. Buchanan has been appointed Safety and Training Advisor at one of the largest skydiving centers in the Northeastern United States, and serves as a Coach Course Director, training other skydivers to be instructors. He holds an FAA Senior Parachute Rigger Certificate, and is an FAA certificated Commercial Pilot with ratings in single and multiengine airplanes, as well as gliders.

Mr. Buchanan made his first jump in 1979 at a small skydiving club near Buffalo, New York. Since then, he has made more than 4500 skydives, and logged more than 65 hours of freefall time.

Introduction

Skydivers have a favorite expression that says, "If riding in an airplane is flying, then sitting in a car is jogging." It's a great adage that neatly defines the uniqueness of human flight. In order to fly, you really need to get out of the airplane. If you do, you will experience the world in a way that you never could have imagined.

I've made more than 4500 skydives, and had the pleasure of flying with thousands of other jumpers, many of them students making their very first skydive. I've always enjoyed teaching because it is so much fun to share my students' emotions as they fly their bodies for the first time. Most people find that first jump is an amazing experience that will be remembered for the rest of their lives.

If you are thinking about making a jump, you probably have a ton of questions. This book is designed to provide answers. It should give you an overview of skydiving, and a great background to help you find and evaluate skydiving schools in the United States. Although every school is unique, I've tried to focus on those things they have in common, and I offer enough information so that you can recognize the differences. This should help you to pick the school that is right for you.

USING THIS BOOK

I've written this book so you can read it in several ways. First, it should be easy to read like a standard book, that is, cover to cover. You will find that each chapter builds on the material in previous chapters, and by the end you will have a complete understanding of skydiving, and how to go about finding a great school.

If you don't have time to read the entire book right away, try focusing on Chapter 1 for a general overview, then flip to Chapter 10 for assistance in selecting a skydiving school. Those two chapters should give you a quick idea of what skydiving instruction is all about, and help you to find a place to learn. When time allows, go back and read the other chapters for a more detailed understanding of the sport, and the skydiving community.

If you are really pressed for time, check the Appendix for a list of common questions that are asked by many students, along with super-fast answers, all presented in an FAQ (Frequently Asked Questions) format. The answers are explained in greater detail in the text, so this section can serve as a quick review after reading all the chapters, or as the beginning point for discussing your skydiving interest with friends and family.

MY FIRST JUMP

I never expected to jump out of an airplane, and I actually began skydiving almost by accident. I was working for a student newspaper at the University of Buffalo when a man came in and offered a free skydive if we would write a story about his parachuting school. It sounded interesting, so I jumped at the opportunity. That's a bad pun, I know, but it has been working for 20 years, so I figured I'd share it one more time. I had a blast on that jump, although I think you can imagine, I was pretty scared. Sadly, I was a poor college student and couldn't afford the cost of continuing training.

I didn't jump again until the following year, when I contacted a different skydiving school and turned the tables a bit, offering to write a story if they would give me a free skydive. This time the school jumped at the opportunity. And once again, I had a blast. I was able to score one more free jump the third year, using the same approach at yet another skydiving school. When I landed following that jump I was hooked, and somehow I found a way to pay for more training. After a handful of jumps the school gave me a chance to assist with packing parachutes for other students, and the money I earned doing that chore helped pay for more skydiving lessons. In due time I earned my first instructional rating and traded the drudgery of packing parachutes for the joy of teaching beginners.

When I made my first jump, the only kind of training available used a static line, an old training method adopted from the military. The parachute was a huge surplus round thing that landed like a ratty paper bag filled with bricks. The airplane was a tiny gutted-out mess. Ugggh. It was all fun, but contemporary training is so much better.

CONTEMPORARY SKYDIVING

If your vision of skydiving is based on watching old movies, or ancient stories of military jumps, you will be surprised to learn that the sport has changed. Today there are many different training methods available, parachutes are rectangular and often land very softly, and most instructors are far better trained than the man who walked into the student newspaper office and offered me my first free skydive.

There are now hundreds of skydiving schools in the United States, and some of them are really great. However, I should point out that the federal government does not regulate the skydiving industry heavily, and some schools don't meet any national standards at all. That makes it really important for you to learn everything you can about skydiving and student training, and then use your skills as a consumer to select the best school for you.

If you are like most people who try skydiving for the first time, you probably just want to make a single jump, and you have no plans for any jumps after the first. Watch out, you may be surprised by the thrill of your skydive and get hooked, just as I did. Keep the future in mind as you read this book. Dream about all the outrageous things skydivers can do, and then imagine yourself in the air making those dreams come true.

I hope this book helps you to understand the thrills of skydiving, and I really hope you enjoy your first jump.

Blue Skies,

Tom Buchanan

Chapter 1

OVERVIEW

*I*t seemed like a good idea a few minutes ago, but right now you are scared. Really scared. You take a small breath, then another, and wonder if stepping out of an airplane is actually such a great idea. Your mind is jammed with random thoughts, but they don't really register. The wind is blasting inside the small plane and it is far colder than you had imagined. The ground is moving by fast, and the field you hope to land in is just a tiny speck a couple of miles below. The harness you are wearing is tight, and you can feel your own heart pounding at about a million beats a second. You try to step closer to the door, but your legs won't cooperate. They feel like rubber. As you think about your stubborn legs and wonder if you should quit now, the instructor moves forward. You realize your harness is attached to his so snugly that you have no choice but to follow along. As you face the door, you want to jump headlong into the emptiness of the sky, but you also want to crawl back inside and say no. Or perhaps you want to scream NO. Or maybe you want to scream YES. It's all so confusing. Nothing is making any sense. Suddenly you hear the instructor's calm voice asking if you are ready to make a tandem skydive. You think you hear yourself answer yes, but your voice doesn't sound real. It sounds like somebody else's. Or maybe it sounds like a movie. You wonder how the instructor can be so relaxed. You wonder if you have the nerve to actually do this. And you wonder if you can ever be as cool as the guy who is about to take you on this amazing ride. Thoughts keep coming, faster and faster, then suddenly your thoughts stop. Your mind stops. It just stops. Nothing happens for what seems like an eternity. You gasp for breath, and then figure out that you have actually jumped, and are now accelerating toward the ground, soon to reach 120 mph. You know you jumped, but you can't remember actually doing it. As you pick up speed, you reach down and touch a ripcord that will open the parachute in less than a minute, and you accidentally touch your instructor's leg. You remember that your harnesses are attached, and that he is wearing the parachute that will save your life. You try to focus on what is happening, but it is all going so fast. Soon, the parachute is opening, and you hear your instructor's voice asking if you want to help steer. Together you fly the parachute and enjoy a unique experience that you don't think you will ever be able to describe. The view is amazing, the air is fresher than you ever imagined, and you feel a sense of relief that you survived the freefall, but you know that you still need to get back on the ground.

Every skydive is an adventure. Every jump is a chance to push yourself and experience life to its fullest. I've been jumping for more than 20 years, and I never tire of the thrill and excitement of stepping outside a speeding airplane miles above the earth. Jumping alone is a kick, and jumping with other skydivers is a blast, but nothing beats jumping with students! Beginners are so loaded with anticipation that when we leave the airplane I can almost taste their adrenalin. The "stoke" they feel after a successful skydive is a rush for them, and for me too!

Photo by Tom Rose

Figure 1.1   Your first skydive will be a memorable experience.

WHAT KIND OF PEOPLE JUMP OUT OF PERFECTLY GOOD AIRPLANES

Skydiving is a quickly evolving sport that is enjoyed by more than 300,000[1] Americans each year. Of those, about 275,000 are students,[2] and most will make only one jump. Some of those beginners will conquer their fear and continue training, looking forward to the day when they can jump alone, or try exciting aerial tricks with other skydivers.

If you are like most beginners, fear and uncertainty will fly hand-in-hand with excitement. Indeed, for many first-time skydivers, one of the main reasons they try the sport is to face their own fears. Others are pulled along by the peer pressure of friends. Most skydiving students have never tried any extreme adventure sport, but some beginners do have experience with other activities such as skiing, rock climbing, or scuba diving. There are several different kinds of skydiving programs, and with a bit of research you can find the school that is just right for your personality.

Skydiving has always attracted a wide range of people. Most skydivers are men, although women are making up a larger percentage of jumpers at many skydiving schools. (See Figure 1.2.) Beginners are drawn from the ranks of engineers, students, business managers, homemakers, pilots, firefighters, truck drivers, the self-employed, and just about every other occupation you can think of. Almost every group imaginable is represented in the sport. When you spend time at a skydiving school, you will find yourself surrounded by a weirdly heterogeneous group of active people brought together by their love of this exciting activity.

You will find skydiving schools located at skydiving centers, a place at an airport where skydivers get together for jumping. Most skydiving centers are located at small airports outside cities or in rural areas. The skydiving center is frequently called a drop zone. Sometimes the skydiving school is

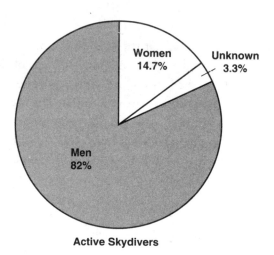

Active Skydivers

Figure 1.2 Most skydivers are men. (*Source: USPA 2001. Graphic by Laura K. Maggio.*)

operated by the same group or company that runs the whole drop zone, but sometimes it is a separate business simply located at the same place.

When you arrive at the skydiving center, you will be directed to the school or training area, and you will be introduced to the instructors who will be responsible for your first jump.

GETTING STARTED

Most beginning students arrive at the skydiving school in a group, to give each other moral support and to share their adventure. Many of those groups are made up of friends who have been thinking of jumping together for several years. Other groups consist of co-workers or members of a social club. Sometimes large groups of beginners can receive a break on the cost of training simply by asking the school for a group discount. If you have assembled a group, it never hurts to ask about special discounts.

Sometimes the groups are large and well organized, but frequently they are simply a small collection of friends out for an afternoon of fun. Occasionally friends will try skydiving to celebrate a birthday or other special occasion. In fact, I once took a student on a tandem skydive to help him propose marriage. In that case we secretly arranged for the groom-to-be to wear a tuxedo under his jumpsuit. I was his tandem instructor, and we left the airplane just seconds before his soon-to-be fiancé made her jump. We opened our parachute lower than his girlfriend and landed first. When his girlfriend landed right in front of us my student had already removed his jumpsuit, then he got down on one knee, presented the ring, and made his proposal. His girlfriend barely had time to catch her wits. It was a unique and romantic moment. Happily, he had picked a great partner, and after just an instant of hesitation she said yes. I have also had students who proposed in the airplane just before we were scheduled to jump. So far the answer has always been yes, usually coupled with a few tears of happiness. It is hard to imagine the anxiety a student must feel when he is planning to make his first skydive and propose marriage at the same time!

Many people begin skydiving in their twenties and thirties, while most experienced skydivers are in their thirties or forties. (See Figure 1.3.) Almost all skydiving centers require students to be at least 18 years old, or the age of legal majority in their state. A few drop zones allow beginners as young as 16, but those schools are very limited and always require written permission from a parent. It is common for a teenager to anxiously await the day when he can skydive, and some can't wait a minute longer than absolutely necessary. Many skydiving centers have hosted beginners on their 18th birthday, and often their parents come along to celebrate.

There are usually no upper limits on the age of a student, and several skydivers have made their first jump well after their

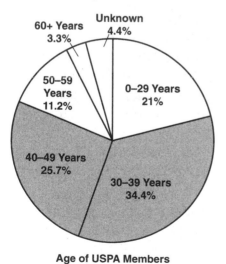

Age of USPA Members

Figure 1.3   Most jumpers are in their thirties or forties. (*Source: USPA 2002. Graphic by Laura K. Maggio.*)

Figure 1.4 Skydiving is enjoyed by people of all ages.

80th birthday. I actually took my own mother on a tandem jump when she was 65, and she loved the experience. In fact, there is a group of skydivers called Skydivers Over Sixty (SOS), with membership restricted to jumpers who are at least 60 years old. When the SOS gang gets together, they always have a blast. Very few active skydivers allow age to keep them on the ground, although older skydivers generally think a bit more about the risk and the added problems of recovering from minor bumps and bruises.

COSTS

Skydiving is not an especially expensive sport in which to participate, but training can be costly. There are several different methods available to make your first jump, and the price will vary between about $125 and $350 depending on the program and school you select. The cost of the first jump course almost always includes initial training, equipment rental, and all airplane expenses. Further training will lead to license qualification and will usually total about $1250 to $2500, including all equipment rental and other jump-related expenses. After graduation from a student training program, the required equipment can be purchased for about $2000 to $6000, or in some cases rental gear may be available at a daily rate. Some schools offer training discounts for large groups and for weekday classes. A few skydiving schools have made arrangements with local colleges to offer a student discount for the first jump course, and a few colleges even have skydiving clubs on campus. A few skydiving schools offer discount package deals that combine the cost of your first jump training with additional jumps.

Once you have a skydiving license and equipment, jumps generally cost as little as $15.00 to $20.00 each. Because of the relatively low cost of a first jump, the sport attracts many college students, but they often find it difficult to cover the advanced training and equipment expenses. Sometimes jobs are available that can help a beginner make extra money at a skydiving center. Some jumpers earn extra income by packing other skydivers' parachutes, after making just a few dozen jumps themselves. Generally, students who are strapped for cash can use their ingenuity and find unusual ways to pay their training bills.

WEATHER ISSUES

You should plan on wearing comfortable clothing that is appropriate for the local weather. Much of the training happens outdoors, and you will probably enjoy relaxing outside while watching other parachutists land. Most skydiving centers will provide a jumpsuit for the actual skydive, but if it is colder than about 50 degrees on the ground you should bring along comfortable tight-fitting gloves and a sweater or heavy sweatshirt. Beginning jumpers should always wear sturdy shoes, sneakers, or boots. Sandals are not recommended for first-timers, although more experienced skydivers may sometimes jump barefoot.

Skydiving is an excellent way to beat oppressive summertime heat, at least for a few brief moments. Temperatures generally decrease as a skydiver climbs above the ground, so when it's a sweltering day on the ground, it will usually be far more comfortable at 12,000 feet. The cool air is so inviting that skydivers occasionally get out of the plane without a jumpsuit, and sometimes even jump without any clothes at all!

While cool temperatures in the sky are a pleasant diversion on a hot day, other kinds of weather can be a problem. Skydivers are not allowed to jump in a cloud, so we are often stuck on the ground when the sky is overcast. Sometimes a group of jumpers will leave the airplane planning to freefall along the side of a tall cloud. When that happens it is really fun because the cloud is a stationary element, and passing it at 120 mph presents an unusual feeling of speed and acceleration. Clouds look soft and cuddly, but they really feel just like thick fog. Sometimes clouds are cool, damp, or wet, but mostly they are just a white mass of nothingness.

Occasionally a skydiver who is falling between the sun and a cloud will look down and see his or her own shadow on the cloud, often surrounded by a circle of white light and a rainbow. The unusual effect is known as the "Specter of the Brocken," or more commonly, simply a "glory." It is very rare, and truly a wonderful sight.

Wind also presents some unique challenges to skydivers. Most solo student parachutes have a maximum forward speed of between 15 and 20 mph. A wind that is blowing faster than a parachute can fly will often push the parachute backwards,

Photo by Dean O'Flaherty

Figure 1.5   Skydivers rarely jump through clouds.

making it difficult to land on a target. For this reason most skydivers stay on the ground when the wind speed is more than 20 mph, and solo skydiving students are often restricted to even lower wind conditions, generally about 14 mph. Tandem landings are actually most comfortable when there is a light breeze, and tandems can often land safely in winds as high as 25 mph.

Turbulence is also a problem for parachutes when near the ground. If you have ever been in a commercial airplane, chances are that you have felt the bumps and swings of turbulence. That effect is always more pronounced in a smaller aircraft, or when flying a gliding parachute. Since a parachute is made of nylon, it doesn't have any ridged surfaces, and it requires stable wind to fly safely. If conditions near the ground are bumpy, the skydiving school will sometimes stop jumping for a while, and wait for the air to settle a bit. Turbulence is invisible, but can frequently be anticipated, or it can be detected by watching trees, wind socks, or other parachutes.

Often students who are on the ground waiting for wind or turbulence to clear will feel frustrated because they are anxious to jump and can't "see" the problem. However, it is always safer to wait for ideal conditions. Turbulence and wind are often strongest in the middle of the day, then ease up near sunset, so often a short wait will yield a safer and more comfortable skydive. Skydivers have a favorite expression: "Sometimes it is better to be on the ground wishing you were in the sky, than in the sky wishing you were on the ground."

LIABILITY AND INSURANCE CONCERNS

Every skydiving center is concerned about the unique risks inherent in the sport, and most will ask students to sign a detailed waiver of liability. The waiver is a standard legal document that will often prevent you from recovering damages in the event of an accident. Although serious accidents are not common, they can happen, and every jumper needs to understand the potential liabilities involved. Most skydiving centers will be happy to provide a copy of their liability waiver in advance, if asked. Most outfitters in the adventure sports industry require participants to sign similar documents. Although the rules and enforceability of an assumption-of-risk waiver vary from state to state, most courts will recognize some form of release in at least some cases. If you are concerned about the content of this important legal document, you should obtain a copy and discuss it with your attorney before visiting the drop zone. The skydiving school may ask for identification when you register and fill out the waiver, so you should bring your driver's license or another form of official identification with you.

Many students carry regular medical insurance as part of a group insurance program at work or school, or as a private policy. Most group medical insurance policies cover injuries while participating in active sports such as skydiving, whitewater rafting, rock climbing, and scuba diving. Some medical policies have specific exclusions, and individual life insurance policies may limit coverage when flying in small planes or skydiving. Insurance agents can sometimes waive those limitations or provide a supplemental policy. It is always a good idea to know what your personal policy covers, and what it does not. Your insurance agent or company human resources office is a good source of information about your current coverage. Very few skydiving schools carry insurance that covers student injuries. If insurance is a concern, you should ask the skydiving school what coverage it has available, and carefully review its policies in advance.

Skydiving injuries are not common, but it is important that anybody who participates in extreme sports is prepared to handle the unexpected costs of injury. An emergency room visit and follow-up care to treat a problem as simple as a sprained ankle, or a broken wrist, can cost hundreds or even thousands of dollars. Those expenses are generally not a huge problem if the injury is covered by insurance, but the expense can cause serious aggravation if money is not available for medical assistance.

WEIGHT LIMITS

Skydiving centers have student weight limits that are based on the Federal Aviation Administration (FAA)-approved certification of their equipment. Regular skydiving parachutes and harnesses are usually certified to hold 254 pounds, including the jumper and the weight of his or her gear. Some equipment is certified by the FAA without a weight limit, while other equipment is certified to a weight limit defined by the manufacturer's testing program. The certification standard includes

Figure 1.6 Tandem equipment is designed for two people.

a bit of a buffer, so the equipment will probably hold more weight, but the approved load is limited to the original certification level.

Tandem equipment, designed for two people, is certified to hold between 500 and 625 pounds, depending on how it was originally tested by the manufacturer. When you make a tandem jump, the limit applies to your weight plus the combined weight of your instructor and the roughly 50-pound tandem rig. Many drop zones impose their own weight limit of about 220 pounds for tandem students. Some tandem programs will allow slightly heavier students to jump by matching them with lighter instructors, but an instructor who is not very big himself will have a difficult time flying with a very heavy student. If you are over the local weight limit most schools will evaluate your physical condition, and they may be willing to take you on a tandem jump if you are physically fit and already lead an active life. Students who are severely overweight and generally inactive will find the harness uncomfortable, and may be at greater risk of injury during the parachute opening and landing. If you think your weight might be a concern, you should mention it to the school when you register for your jump. Some schools that are willing to take heavy students may charge a few extra dollars as a way to compensate their instructors for the added workload.

There is no problem with being underweight, and overall physical size does not matter much either. Skydiving is a sport that can be enjoyed by people as short as professional horse jockeys, and as tall as basketball players.

MEDICAL ISSUES

You should be in generally good health when you try skydiving. A first jump is stressful and probably not recommended for people with advanced heart disease, or other serious medical problems. Issues like diabetes, asthma, seizures, and other special medical conditions should be discussed with a doctor before trying skydiving.[3] Although these conditions are often not exclusionary, there are special concerns that need to be discussed with the skydiving school. If you are concerned about a specific medical condition, it is always a good idea to contact several schools and seek a variety of opinions. Most schools will appreciate your willingness to discuss your concern, and will be happy to share their experience and insight.

Tandem skydiving (jumping while attached to an instructor) is a great way to introduce beginners with disabilities to the sport. Many paraplegic and quadriplegic jumpers have safely experienced tandem skydiving. Blind and hearing-impaired people have tried skydiving with special help from experienced instructors. There is even an eight-person team of experienced skydivers in California who call themselves Pieces of Eight, because each is jumping in spite of an amputated limb. Students with special needs generally require a bit of extra support, and that should be coordinated with a skydiving school in advance. Most drop zones will try to accommodate jumpers with special physical needs, but some will refer a student to a school that already has experience with the specific issue. If you anticipate special needs, it is a great idea to call the skydiving school in advance to be sure a local instructor has the required background and understanding of the issues involved. The Uninsured Relative Workshop, a skydiving equipment company, has produced a guidebook called *Skydiving with Wheelchair Dependent Persons*, which offers tips for making these unique jumps. Most skydiving schools will be happy to obtain a copy of this booklet for you at no charge. For some students, skydiving with a physical challenge is a unique

Figure 1.7   Many skydiving schools can accommodate students with special needs.

event that adds a new feeling of independence. Other folks with special physical conditions are accustomed to breaking through any barriers that restrict their enjoyment of life, and they often see skydiving as just one more experience in an active lifestyle.

Gasping for Breath

One of the strangest medical issues that students occasionally face is an inability to breathe in freefall. Although this is a rare problem, many students wonder and worry about it. The problem will sometimes develop early on a skydive, leaving the student gasping for breath and frightened of suffocation. Oddly, there is nothing wrong, and no physical reason for the problem, but nevertheless it is very real. Fortunately, the problem is easy to solve. The cause of breathing difficulty is usually tension in the chest because the student is inadvertently holding his breath, or he simply forgets to breathe. That simplistic explanation might sound silly right now, but if you are falling through 10,000 feet and unable to breathe you won't think it is silly at all. The solution is simply to relax and breathe—easy to say, tougher to do. One of the key exercises that instructors often suggest to their students is to consciously breathe through the nose, or in through the nose and out through the mouth. The specific process isn't all that important, but just thinking about breathing in that way is usually enough to clear up the problem. In any case, breathing problems are always limited to the first jump or two, and the issue is easy to solve.

Some students are convinced that there is magic to breathing in freefall, but there isn't. We all know people who ride motorcycles at highway speeds, and they have no problem breathing high-speed air at 60 mph; and breathing at 120 mph is really no more challenging.

Nausea

Some beginners are worried about feeling nauseous after the parachute opens. Although this does happen occasionally, it is a rare problem usually caused by anxiety or motion sickness. Sometimes a misadjusted harness can push on a pressure point and make a tandem jumper feel sick. Frequently, the best way to deal with the feeling of nausea is to engage your mind in something else, such as gently flying the parachute. It also helps to take a few deep breaths, and to hold your head pretty still for a moment while stopping any turns the parachute may be making. The deep breaths will get oxygen to your brain, and holding your head still will allow your ears to stabilize. If you are jumping with an instructor on a tandem skydive and you feel sick, you should mention your upset stomach so the instructor can slow the parachute down and minimize radical maneuvers. Some students try to control the threat of nausea by not eating before a jump, but this can often make you weak, and upset your stomach more than the jump. If you can, you should try to eat a normal meal a few hours before your jump, but avoid greasy and spicy foods. Soft foods such as simple sandwiches, cookies, or popcorn can often help to keep your stomach settled. Feelings of nausea are unusual in the skydiving world, and are limited to beginner skydivers. Often people who have trouble with motion sickness on roller coasters will not experience the problem while skydiving. If you think you might feel sick, it should help you to know that the problem is both rare and easy to solve, and almost always limited to the first few jumps.

Ear Congestion and Sinus Blockages

You have almost certainly experienced the effects of pressure changes in a swimming pool, an elevator, or in a commercial airplane. Air pressure changes on a skydive too, and it is usually easy to deal with. You will probably feel the pressure change in your ears on the way up in the plane, and then when descending. Most people can easily handle the change on the way up because it is so gradual, but some students have a hard time adjusting to the rapid pressure change in freefall. The pressure change in freefall is not really that great—it is often less than the pressure change you experience when diving to the bottom of a deep swimming pool or lake—but it can be a problem for some students if it is not dealt with quickly.

If you have a hard time with pressure changes in elevators or airplanes, you should be prepared to equalize your ear pressure just after opening. If you feel the squeeze of pressure build up when the parachute opens, you should try to equalize by swallowing. If that doesn't work, close your mouth and pinch your nose, then try to exhale gently through your blocked nose. This technique will push a small amount of air toward your ear canal and equalize the pressure. It sometimes helps to wiggle your jaw back and forth as you push air against your blocked nose. You may need to repeat this as the parachute descends, and then again on the ground. The trick to controlling pressure buildup is to equalize pressure early and often. After a few skydives, the process of equalizing will become almost automatic. Most students don't have a problem with pressure equalization, but some are concerned about it. Knowing how to solve the problem will often help alleviate anxiety.

Serious ear blockages can occur if you make a skydive with congested sinuses. You should avoid skydiving if you have a cold or are experiencing congestion from serious allergies. Some decongestants may help keep your ears clear, but they may also have side effects such as drowsiness, so they should be avoided unless you have discussed their use with a doctor.

SENSORY OVERLOAD

Often a beginning student will be so overwhelmed after leaving the airplane that he or she won't remember much of anything at all about the first few seconds of the jump. Skydiving instructors call this problem "sensory overload," and it affects just about every beginner to some degree. Often the experience of leaving an airplane presents your brain with so much new information that it just can't process everything quickly enough, so the brain closes itself off to the world for a brief instant while it figures out what is happening. Most students recover on their own in just a couple of seconds, often not even realizing they have a limited memory of that brief instant. Often a quick shake from the instructor or the tug of a parachute opening is enough to jump-start your brain. Over the next several skydives the problem will lessen, and eventually it will go away entirely.

One of the best ways to counter the effect of sensory overload is to have a specific task to accomplish right after leaving the plane. Something like checking for the location of a ripcord,

Photo courtesy of Archway Skydiving Center

Figure 1.8   Sensory overload will often make it difficult to remember the exit and the first few seconds of a jump.

or watching the airplane fly away, will often be enough to provide a mental focus point and lessen the effect of overload. It also helps to have a clear sense of what you expect to happen on the jump, and at least a general idea of what it will feel like. Knowing what is likely to happen will make it easy for your mind to process all the new information based on established expectations. You can help your brain to do its job, and reduce the effect of sensory overload, by learning as much as possible about the jump in advance. As you gain experience in skydiving your mind will have a much easier time processing information, and the period of overload will diminish.

NEED FOR SPEED

Many skydivers enjoy the special rush of flying through the air at high speed. Skydiving has been called the fastest nonmotorized sport because experienced jumpers can freefall as fast as 300 mph, or slower than 40 mph, depending on altitude, body position, and the special equipment they carry. Most students fall at speeds in the range of about 110 to 130 mph, or about 175 feet every second. That is faster than most people ever travel on the ground, and the wind blast is amazing. Usually that wind is the only thing that creates a sense of speed. While you are in freefall the ground is so far away that it does not look like it is moving at all, so speed is hard to detect visually.

Many skydiving centers offer students a chance to have their skydive videotaped and photographed by a photographer who actually jumps with them. This is a great way to document and remember the experience, and watching the photographer when your parachute opens will really give you a rush of speed. When your parachute begins to open, you will immediately start to slow down, but the photographer who will be right in front of you, will still be moving at about 120 mph, so you will have a chance to see him fall away super-fast.

If you have video coverage on your jump and can remember to watch the photographer at pull time, you will be rewarded with a stunning sight of breathtaking speed.

One of the common misconceptions about parachutes is that a jumper goes up when the parachute opens. This is generally based on watching film and video of jumpers at opening time, and they often do look like they are going up. Actually, the effect is an illusion created by the perspective of the photographer, who is continuing to fall at the same speed while the jumper under the opening parachute is slowing down. When your parachute opens, you always continue to descend for a short while, but at a diminishing speed.

Once the parachute opens, your descent rate will be closer to 1000 feet per minute, or just about 11 mph. Modern parachutes generally fly forward at about 15 to 30 mph, and can be easily controlled. A parachute is really a glider that flies very much like an airplane without the motor. Parachutes are

Figure 1.9 Watching a freefall photographer when your parachute opens will give you a sense of speed.

controlled by two toggles, small loops of nylon webbing attached to lines, which are in turn attached to the back of the parachute. Pulling down on the right toggle steers the parachute to the right, and pulling down on the left toggle steers the parachute to the left. When both toggles are pulled at the same time the parachute slows its forward speed and descent, an ideal way to make landings soft and comfortable.

Many older skydivers remember when students were trained with old military-style round parachutes that slammed a jumper into the ground every time. Contemporary sport parachutes are rectangular in shape, and they make landings much more comfortable and controllable. It is really easy to steer a student parachute to a landing right back at the school, and often those landings will be made standing up. In many ways, basic parachute flight is simpler than driving a car, and the instruction you receive as part of your jump training will make the parachute ride as much fun as the freefall. You may

Photo courtesy of Brentfinley.com

Figure 1.10   Skydiver Mary Traub flies her parachute over San Carlos, Mexico. Control is maintained with steering toggles.

even find the parachute ride inspiring, and more enjoyable than the high-speed part of the jump!

EMERGENCIES

Some students worry that they may faint or pass out, either because of excitement or fear. This is extremely uncommon, but it can happen, and your tanden instructor has been well trained to land the parachute without your assistance. Although landing with an unconscious student is rare, it is generally not a problem, and most students recover shortly after reaching the ground.

If you are like most beginners, you hope that your parachute will open properly, but you probably have some concerns about what will happen if it doesn't. All parachutists are required to have two parachutes for every jump, a main and reserve, and every skydiving center complies with this regulation. Frequently the reserve is just like the main, with a few subtle but important differences. If the first parachute doesn't work properly, the reserve is always available. Most jumpers use a single rig that holds the main and reserve parachutes on their back. The main is in the bottom of that dual container, the reserve is in the top. The reserve parachute must be inspected and repacked by a federally licensed parachute rigger at least every 120 days, so it is very well maintained and reliable. Students making a tandem skydive can rely on their instructor to deal with any parachute malfunctions, while students making a solo first jump will be trained to handle any foreseeable problems on their own. In most cases your reserve parachute will be equipped with a special computer called an Automatic Activation Device (AAD) that measures your descent rate and altitude. The AAD is designed to open your reserve automatically if you are experiencing a problem with your main parachute, even if you don't open it yourself.

Photo by Tom Rose

Figure 1.11   Skydiving is a lot of fun.

Skydivers are taught to follow emergency procedures quickly, and not to rely on this device, but it sure is great to know that an advanced computer is ready to help out.

Skydiving is an exciting sport that is unlike anything else you have ever tried. It can be done relatively safely with quality instruction, but if handled recklessly, it can become a very dangerous activity. A great skydiving school will provide high-quality equipment, well-maintained airplanes for the short ride to the jump altitude, and a well-qualified instructor. A top-notch instructor will not only help you to skydive safely, but will also serve as an expert guide throughout your jumping adventure.

Chapter 2

TRAINING METHODS

Skydiving has undergone remarkable changes over the last 20 years. The airplanes available for skydiving are larger, faster, and safer than ever before, and equipment and training have improved dramatically. Contemporary training programs offer a variety of ways to make your first skydive, and after the first jump there are many advanced training options. You may decide to make just one jump, or you may be hooked by the excitement and choose to continue your skydiving training.

TANDEM JUMPS

If you are like most students, your first skydive will be a tandem jump. Most schools offer tandem training, and many will require a tandem jump before you can make any kind of solo jump. Modern tandem jumping actually began in the early 1980s, and was operated as an experimental program until July 2001, when the Federal Aviation Administration (FAA) formally recognized tandem as a regular part of skydiving.[4] Tandem skydiving may be the newest way to make a first skydive, but it has quickly become the most popular training program.

Figure 2.1   Your instructor will make the needed connections.

Tandem jumping allows you and your instructor to jump together, providing you with direct supervision and support throughout the jump. To make tandem skydiving work, the tandem student is fitted with a special harness. The instructor wears his own harness, which has a main and a reserve parachute on the back. When it is almost time to jump, the instructor connects the two harnesses together using special hardware. You will be in the front position, and the instructor will be in the back, with both of you facing forward. (See Figure 2.1.)

Training

Since you will be attached to a tandem instructor, very little training is required for a tandem first jump. Most of the impor-

Figure 2.2 You may enjoy the airplane ride as much as the skydive.

tant instruction can be accomplished in as little as 15 minutes. Even advanced training can be completed with about 30–45 minutes of ground time. Because the classroom time is so limited, you can often join your friends for a tandem jump in the morning, and then have the rest of the day available to celebrate your accomplishment.

Many skydiving schools take great pride in teaching students far more than they will need to know for a single tandem jump. Schools with a reputation for first-class training often teach you how to turn in freefall, how to read an altimeter, how and when to pull the ripcord, and how to fly the parachute. Many of these training centers will even encourage you to pull the ripcord yourself, and will often let you fly the parachute on your own. Some schools will even let you help with the landing. Of course, if you would rather not pull the ripcord or join in the flying, the tandem instructor is always right there to handle everything.

Some beginners are initially reluctant to take responsibility for anything on a tandem jump, but after a little bit of encouragement most are happy to become part of the excitement. Some schools teach their students that if they pull the ripcord on a tandem skydive, they can tell their friends that they "saved their instructor's life," and that's a really cool boast! Other skydiving centers view a tandem jump as simply a joy ride and provide their students with a bare minimum of training. These skydiving centers will generally expect nothing from the student, and will instead have the instructor handle everything on the jump. Often students trained at this kind of school will not have access to a ripcord, or any training on how to fly or land a parachute. Some beginners seem to like the idea of letting an experienced instructor handle the entire skydive, and prefer a program with a minimum amount of training.

You will probably find that tandem jumping is the most relaxing way to experience skydiving because you will not have any real concerns about what to do if something goes wrong. Because the instructor is physically attached to you, he is able to deal with parachute malfunctions or anything else that might happen. Another great advantage of tandem skydiving is that the instructor is close enough to talk with you. Although it is difficult to communicate in freefall, it is possible to talk back and forth once the parachute opens, and that makes tandem jumps great for teaching advanced flying skills.

Altitude and Speed

Tandem skydives are generally made from altitudes between 8000 and 14,000 feet above the ground, and offer freefall times between 20 and 50 seconds. Shortly after exiting the airplane, a tandem instructor will deploy a small "drogue chute," which will trail behind the pair for the rest of the freefall. The drogue is actually a tiny round parachute that will stop your acceleration and hold freefall speed to about 120 mph, a normal speed for a single jumper. Without a drogue the extra weight of the tandem pair would cause them to accelerate to 180 mph. That extra speed does not offer a very good training experience, and parachute openings at 180 mph are uncomfortable for the jumpers, and taxing on the equipment. The drogue is initially stowed in a small pocket on the bottom of the parachute rig, where the instructor can easily reach it. Often a student will not feel any difference after the drogue has been deployed. (See Figure 2.3.)

Photo by Jim Stahl

Figure 2.3 A Twin Otter airplane descends after releasing a tandem. Tandem freefall speed is controlled with a drogue.

As the parachute opens, the drogue will collapse and trail behind the open parachute. If you look at photographs of tandem parachutes you will often see the drogue behind the instructor, or perhaps you will see the drogue bridle line extending out of the frame.

Once the parachute is open, the ride down will take 3 to 5 minutes. If you have a chance to help out with steering the parachute, you will quickly see how easy it is to control the flight, and make the parachute go wherever you want it to go. Tandem landings are usually accomplished either standing up, or sliding along the ground on your legs and buttocks. Your instructor will decide what landing method will work best for your jump based on the wind, and the performance of the parachute.

After you have made one tandem jump you may be so stoked that you want to get right back in the air again. Many skydiving centers will be happy to help you make as many tandem skydives as you like. On future jumps you will learn more about controlling your body in freefall, and may be able to try advanced maneuvers such as flips. You might even find the parachute ride so enjoyable that you will ask your tandem instructor to open a bit higher than normal, and spend more time just flying around the countryside.

Your first tandem jump will generally cost between $150 and $250. Most skydiving centers offer advanced training after your tandem jump, and some offer discounts for the jumps after your first.

STATIC LINE JUMPS

The oldest form of skydiving instruction is static line training. The static line program got its start with military jumping in the early 1900s, and is still used as an effective way to deploy large numbers of troops into battle quickly. Civilians later adopted the static line method, and for a long while it was the only way of training skydivers. In recent years tandem training has overtaken static line as the most common way to make a first jump, but static line is still very popular, especially in rural areas, where other forms of training may not be available.

If you make a static line jump, you will wear your own parachute rig with both a main and a reserve parachute. The main parachute system is attached directly to the airplane using a very strong line called a static line. When you leave the airplane, the static line becomes taut and automatically pulls a pin that holds the main parachute container closed, and then helps stretch out the parachute. Once the parachute

Photo courtesy of Archway Skydiving Center

Figure 2.4 A static line jump can be a lot of fun.

begins opening, the static line separates, and you will continue falling away until your parachute is fully open and flying. Since the static line is still attached to the airplane, it will remain behind, and your instructor will need to pull it back inside the plane before he jumps himself. Most static lines are about 8 to 12 feet long, so the parachute opens very quickly.

A Different Way

Some skydiving schools have modified their static line program to use a conventional student parachute without an actual static line. These programs use a pilot chute, a small round parachute-like drag device, to open the main parachute. The pilot chute is attached to a length of nylon bridle, which is in turn attached to the pin, and then to the actual main parachute. Using this system, instead of connecting the static line directly to the airplane, the instructor holds onto the pilot chute until you jump. At that point he or she throws the pilot chute behind you. As the pilot chute fills with air, it serves as a giant anchor in the sky, and pulls the pin to begin opening the parachute. Experienced skydivers use a pilot chute to open their main parachutes rather than a traditional ripcord or static line, so using this same kind of system for student jumps makes it easier and less expensive for a skydiving school to operate. This kind of student training is called instructor-assisted deployment (IAD), and is so similar to a static line that many schools use the two systems almost interchangeably. If you make your first skydive at a school that uses the IAD method, you should know that your instructor is directly responsible for initiating your parachute opening by throwing the pilot chute behind you. In a conventional static line program the opening happens automatically as you fall away from the airplane. The difference is subtle, but important to understand. In all other ways static line and IAD programs are the same.

Training

Since you will not have an instructor jumping with you on a static line jump, it is critical that you are able to fly the parachute by yourself, and know what to do if something goes wrong. The training you will receive before making a static line jump will cover such things as finding the airport, exit stability, flying the parachute, landing, and how to deal with any parachute malfunction that you might experience. Often a school will provide you with a radio receiver, and an instructor on the ground will offer directions to help you steer the parachute. He or she will also provide radio assistance with the landing.

Initial static line jumps are usually made from 3000 to 3500 feet above the ground. Since the parachute opens right away, no freefall is involved. The parachute ride will be a

Figure 2-5   An instructor often provides parachute guidance by radio.

breathtaking experience and should last about 3 minutes. If you steer correctly, you will probably land in a big field back at the drop zone.

A static line jump is far more than just a ride, and that is one of the reasons this training method has remained popular with students and skydiving schools. Since you are flying on your own, you must be willing and able to take charge, and you will be responsible for everything that happens on the jump. Some students feel this is too much responsibility and prefer another way to experience the sport. For many students, however, nothing is more fun than landing on their own at the end of a great static line jump.

After making your first static line jump, you can continue training in a complete static line program. If you make your first jump using a static line, most schools will expect you to make a minimum of four more static line jumps before you will be allowed to try freefall. Those five jumps will give you time to fly the parachute several times, and you will begin to practice pulling a ripcord, or perhaps practice throwing a pilot chute to open the parachute. When you have made five static line jumps and demonstrated adequate control, you will be able to leave the airplane and open your own parachute immediately by making an exciting advanced jump called a "clear and pull." That is an amazing moment of accomplishment when you have become fully capable of jumping from the airplane and opening the parachute on your own. As your experience builds, your instructor will take you to higher altitudes, and you will be able to make longer freefalls. Those jumps will allow you to build advanced flying skills while under the supervision of an instructor.

A typical static line first jump course will take about 5 to 6 hours, and you can usually make your first jump on the same day you complete the ground training. A static line ground course with all the required training, equipment rental, and jump fees will generally cost about \$125 to \$175. Training for subsequent jumps will generally take less than an hour, and cost only about \$40 to \$75 for each jump.

ACCELERATED FREEFALL

Accelerated Freefall (AFF), a training program developed by the United States Parachute Association (USPA), combines the thrill of a solo parachute ride with the excitement of freefall. The AFF training method is the most aggressive way to learn skydiving and build advanced flying skills. The AFF program is designed to have you making solo freefall skydives without an instructor in as few as seven jumps.

Relatively few students begin their training with an AFF jump, mostly because it is expensive, and it demands so much of the student. The Accelerated Freefall program was initially developed in the late 1970s and was adopted as a national training program in 1981. The core structure of the program has remained pretty much unchanged since then. Former President George Bush became the world's most famous skydiving student when he made his first AFF-style jump in 1997 at the age of 72.

How It Works

If you choose to make your first skydive under the AFF program, you will be equipped with a student parachute rig that has a main and a reserve parachute, and either a conventional ripcord or a pilot chute mounted in a pocket of the rig. You will probably jump out of the airplane between 9000 and 13,000 feet, and will experience about 30 to 50 seconds of freefall before you open your own parachute. When you leave the airplane, two very experienced instructors will actually be holding on to you, and they will not let go throughout the freefall. When you reach about 5500 feet you will pull your own ripcord, or throw your own pilot chute. As you do that, your parachute will open and your instructors will fall away. You will experience the rush of being all alone under your own parachute about a mile above the ground. The parachute ride will take about 4 to 5 minutes, and you will be guided by an instructor on the ground who will direct you by radio.

Since you will be experiencing freefall, you will need extensive training in a freefall body position called an "arch." This position will allow you to fly comfortably on your belly with maximum control. The arch is a foundational skydiving skill that is covered in other training methods, but it is critical in the AFF program, so you will generally spend a lot of time perfecting this basic freefall position. You will also need to know how to communicate with your instructors, and since there is so much wind noise, you will need to learn special hand signals. Sometimes you may see skydivers in movies who are able to talk calmly while in freefall, but that is strictly a Hollywood invention. Skydiving sign language is really the only way to communicate at 120 mph. (See Figure 2.6.) You will also need to learn an easy recovery technique in case you begin tumbling, and what to do if your instructors are not able to hold onto you in freefall. AFF students also learn to pull their own ripcord or throw a pilot chute on the very first jump, and since that skill is so important, you will find yourself practicing the ripcord pull several times in freefall as soon as you leave the airplane.

PULL!   **Legs in (6 inches)**   **OK!**

Photo by Mike Lanfor

Figure 2.6 Instructors can communicate with you by using freefall hand signals.

Once your parachute opens the instructors will fall away, so you will need to know how to find the airport, how to steer your parachute for a pinpoint landing on the drop zone, and what to do if the parachute malfunctions or something goes wrong. An AFF skydive is a very demanding jump and it is easy to lose track of time, so you will be taught how to maintain altitude awareness in that active environment, and you will probably be taught some simple relaxation and concentration skills that will help you focus on the critical elements of your training.

Accelerated Freefall training is far more comprehensive than static line training, and it takes a bit longer to complete the classroom sessions. This training method requires very experienced instructors who have earned an elite rating in AFF-style harness-hold instruction. AFF instructors are usually the best-trained and best-tested instructors on the drop zone.

Generally, an AFF ground course will last a minimum of about 6 hours. Most students are able to make their first jump on the same day they do the ground training. A typical AFF first-jump ground course with all required training, equipment rental, and jump expenses will generally cost $250 to $325. Subsequent jumps start with about 30 to 60 minutes of ground training, require either one or two instructors to accompany you in freefall, and generally cost $125 to $200. The cost tends to be high for the early jumps because two instructors accompany you in freefall. After a few jumps you will only have one instructor along in freefall, and the cost of the training will usually go down.

Once you have mastered the basics of freefall stability, altitude awareness, parachute deployment, and control, you will be taught things such as freefall turns, flips, rolls, and how to fly together with other people. The AFF program is a great way to learn these advanced skills because you are flying your

Figure 2.7   AFF ground training is comprehensive.

own body, and there will always be an instructor in the air with you to provide training and feedback.

MAKING MORE JUMPS

If you make your first skydive in the tandem program, you may decide to jump again right away. Many tandem students find they enjoy making two or three jumps in a day, but after about the third jump, they tend to become overwhelmed. You will probably be happy with just one or two tandem jumps each day until you have more experience.

If you make your first jump in the static line or AFF program, you will probably jump immediately following the long ground school. Some students can handle a second static line or AFF jump on that day, but most feel that one is enough. When you come back for additional jumps, there is less ground training, and you will be able to make two or three jumps in a day, assuming the weather is fine and the skydiving school has the staff to handle your interest.

VIDEO TRAINING

Video coverage of your skydive is a great addition because it allows you and your instructor to review each part of the jump in detail while relaxing on the ground. Students and instructors like video because it can really help you visualize what you are doing, and makes the learning process much more efficient. Many schools offer video and still photography of your jumps for an additional charge. Some schools have a camera mounted outside the airplane to shoot your exit, while others may send a photographer along in freefall with you. The freefall photographer will often have a still camera and a video camera mounted to his or her helmet. The video camera will be turned on just before you exit the airplane. Some schools will edit your video right away and hand it to you, along with an exposed roll of film. Other schools will mail the finished tape, DVD, or photographs to you later. Videotape shot specifically for training use is generally available within minutes of landing, and is a big help as your instructor reviews the learning objectives of your skydive. Even if you don't plan on continuing with your skydiving training, you will find that photo coverage is a great way to remember the jump, and to share the excitement with your friends.

FREEFALL SIMULATORS

Skydiving can be a very difficult sport to learn because each jump happens so fast, and even with hand signals it can be tough to communicate with your instructor while in freefall. Skydiving also requires several different and unusual skills such as three-dimensional freefall body control, heading control, time and altitude awareness, parachute flight, and outstanding judgment. It is difficult to teach all these topics at the same time, and especially tough to practice them in tiny 1-minute blocks.

Fortunately, there have been rapid advances in custom simulation programs that can help you train for skydiving, and you won't even need an airplane. We can now isolate specific skills and work on them without ever leaving the ground. Simulation allows you to spend far more time gaining important experience under more comfortable conditions, with far less stress than on an actual skydive.

Vertical wind tunnels are amazing simulators that allow you to fly your body without jumping from an airplane. These massive tunnels have a powerful motor in the floor or ceiling that creates a high-speed cushion of air. That air column feels just like freefall, and allows you to learn stability and practice body control. These tunnels are so advanced that every world-class skydiving team now schedules tunnel training time to improve their own skills. Coaches and skydiving instructors often book time in these tunnels for basic student training, and to teach advanced beginners. Tunnels also allow for lots of quick training sessions, with a chance to talk with your instructor immediately after practicing a skill. Wind tunnels offer the feel of freefall without the hassle of packing a parachute, or the anxiety and risk of making a real jump. (See Figure 2.8.)

Tunnels also offer the opportunity for anyone to feel what skydiving is like regardless of their age, so kids can build their freefall skills before they are old enough to make a jump. Even people who are terrified of airplanes and will never make a skydive can experience this form of human flight.

There are three kinds of wind tunnels. Open-air machines have a giant fan at the bottom of a deck. You push your body into the air and then fly on an invisible vertical column of moving wind without any walls. These machines are

Figure 2.8 Vertical wind tunnels provide great training.

often located at outdoor amusement parks or festivals and offer a quick sense of what freefall is like. The air in open wind machines is somewhat turbulent, and you must wear a big baggy jumpsuit, so it is difficult to do advanced skydiving training. In spite of the drawbacks, the availability of these machines makes them useful for an introductory experience, and they offer a good general introduction to freefall body dynamics without the stress of making a real skydive.

Flyaway Indoor Skydiving

Closed wind tunnels that have a fan on the bottom are a giant improvement over open-air machines. These are true vertical tunnels inside a circular building, and they provide much more stable air for skydiving training. The fan is usually a huge airplane-type propeller that blasts air up through a series of baffles and nets that smooth the flow, and protect you from falling into the fan. Two of these enclosed wind tunnels are run by a company called Flyaway Indoor Skydiving, located in Pigeon Forge, Tennessee, and Las Vegas, Nevada. Both sites offer great introductory rides for people who have never made a skydive. Both tunnels also have local skydiving instructors available on staff to provide more advanced training and help you master body control. The first time you visit one of these tunnels, you will receive a safety briefing that includes an overview of tunnel flight and some pointers that will help you stay in the vertical air column. Each flyer will have just a couple of minutes per session to fly, with additional sessions available as needed. The walls in the Flyaway tunnels are padded for safety, and there are cushions on the floor around the edge of the air column. Flyaway offers several different training options and block time packages. These packages can cost just a few dollars, or several hundred dollars, depending on how much tunnel time you buy. Video coverage is also available to help you master advanced skills, or as a souvenir to take home.

Flyaway was the first company to build an indoor wind tunnel specifically for skydivers, opening the Pigeon Forge facility in 1982. The U.S. Army made extensive use of the original tunnel and was so happy with their success in training military skydivers that they built their own tunnel at Fort Bragg several years later.

SkyVenture Wind Tunnels

The most powerful civilian wind tunnels in the United States are owned by a company called SkyVenture. The closed-tunnel design used by this company features several giant fans on the top of the tunnel that pull air up, rather than blowing it from the bottom.

SkyVenture opened its tunnel in Orlando, Florida, in 1997. It quickly became a favorite of competitive skydivers and instructors, as well as tourists who have never made a skydive. The air column in the Orlando SkyVenture tunnel is 12 feet wide and extends all the way to the padded walls. Since the wind extends all the way to the edge it is very smooth for flying, and there is no danger of "falling off" the column, making flight much more realistic than in most other tunnels.

SkyVenture has added a second tunnel in Perris, California, and has plans to build more tunnels in Arizona, New Jersey, Nevada, and Colorado. SkyVenture also has several tunnels in other parts of the world, including some that are designed exclusively for military training. (See Figure 2.9.)

All SkyVenture tunnels have observation rooms and integrated video systems. You can visit a SkyVenture tunnel and fly your body after a short 15-minute classroom session,

Photo courtesy of SkyVenture Inc.

Figure 2.9 SkyVenture operates several tunnels around the world, including this one in Orlando, Florida.

then watch a video of your performance. SkyVenture has experienced coaches and instructors available to help beginners master the indoor skydiving experience, as well as to assist experienced skydivers who want to learn new skills. SkyVenture offers several training packages that can help you learn the basics of tunnel flight and develop advanced skydiving skills. These packages cost from a few dollars to several hundred dollars, depending on how much tunnel time you need, and the level of instruction desired. Advanced training is available through close relationships established with local skydiving schools. Some of these schools offer first-jump training programs that combine tunnel time with real skydives.

If you have a chance to visit one of these tunnels, you will probably see some amazing flyers who have spent lots of time in artificial wind columns but have very little skydiving experience. Some of these "tunnel rats" began training as young kids, and will someday become real skydiving students. With all their wind tunnel experience they will really impress their instructors on their first jump.

Wind tunnels have been available for about 20 years, but most skydivers did not recognize the enormous training value they offered until recently. As more skydivers began using tunnels, the training value became apparent, and the popularity of the tunnels increased.

PARACHUTE SIMULATORS

Many years ago the U.S. Forest Service decided it needed a better way to train skydiving firefighters who jump into forest fires. The Forest Service contracted with a company to design a parachute simulator that allows a jumper to experience a virtual reality parachute flight without ever leaving the ground. The system features a special pair of virtual-reality goggles, a training harness with controls, and a sophisticated computer that generates moving graphics. This training device was later adapted for use in military operations, and has recently been introduced to the civilian world. When Former President George Bush trained for his 1997 skydive, he used a virtual-reality parachute simulator. (See Figure 2.10.)

A few civilian skydiving schools are already using parachute simulators. If you train at one of these schools you may have a chance to experience this three-dimensional computer system. The device allows you to see a computer-generated view of a parachute malfunction, and then respond by pulling the correct emergency handles on your harness. If you make a mistake it will allow you to repeat the lesson as many times as it takes you to master the skill. The simulator will also help you learn to navigate your parachute in different kinds of wind, and it will provide training in landing your parachute.

Figure 2.10 Some skydiving centers use a virtual-reality training device called Sport ParaSim.

Skydiving centers that have been using these simulators report they have been able to improve student learning and reduce landing injuries. Students seem to like the addition of computers to skydiving instruction, and find the training allows them to practice critical safety tasks without any stress. Parachute simulators are relatively inexpensive to purchase and operate, and take up very little space. While parachute simulators are still very new, they should become more common as the technology improves and is adopted by the civilian skydiving community.

HYBRID TRAINING PROGRAMS

Not long ago you would have been limited to learning skydiving in just one of the basic three training programs. A recent trend among skydiving centers is to take parts of each core training program and combine them to create custom offerings. Drop zones are currently offering locally developed programs advertised as IAF, AFP, TAF, ATP, ASI, and many others. Most of these programs start with at least one tandem skydive, and then progress into a classic AFF-style harness-hold program or static line-type jumps. Some of the hybrid programs use wind tunnels to build initial experience, and then follow up with tandem or AFF-style harness-hold jumps. Often these programs are referred to as part of an integrated progression and feature a comprehensive ground school (CGS), or advanced ground school (AGS) as part of the training. You may hear some of these acronyms as you research skydiving schools, and may need a super-deluxe-mega-high-power decoder ring to figure out all the offerings. Yet, when you examine skydiving programs they can be broken down into the three basic styles of training: tandem, static line, and AFF (harness-hold). Each style of training has unique benefits, and one of them will probably appeal to you.

Many students and instructors are finding that a tandem jump provides a great introduction to the skydiving environment. After that first jump, additional training can be selected from the other programs to custom-tailor a training plan to your needs. Tandem skydiving offers a direct, hands-on freefall training session coupled with a supervised parachute ride and limited stress. Static line offers the advantage of lots of solo parachute rides from relatively low altitudes, an important consideration in cloudy climates. Accelerated Freefall offers real solo freefall training in a much more intense program. Wind tunnels and parachute simulators allow for the training of isolated skills, and lots of repetition at relatively low cost. It is easy to combine parts of these programs to suit just about any student.

Your skydiving school will probably have a detailed training progression available to help you master advanced skills after your first jump. There are many different ways to combine skills and training jumps from the tandem, Accelerated Freefall, and static line programs.

The first part of any training program is designed to build freefall stability, altitude awareness, and parachute-handling skills so that you will be able to jump without an instructor. The second part of the training adds basic flight skills and maneuvers such as turns and loops. The third part of the training builds more advanced skills, so you can fly with other jumpers. It should take about 20 jumps to develop enough skill to skydive with very experienced jumpers who are not instructors. After about 40 or 50 jumps you will have the skills to fly with just about anybody else who has completed a training program, as long as the school approves your flying partners.

Most programs are usually designed around specific freefall and parachute skills that have been assembled into levels or categories. Each jump is tailored to help you master the specific skills in one category before you move on to the next category. Every school should have a formatted syllabus that details exactly what you will learn on each jump of the progression.

One Training Option

A typical hybrid program starts with about 30 minutes of ground training, followed by your first tandem jump. This skydive is usually designed as an introduction to freefall and parachute flight, and as a way for you to experience the sport with limited stress and responsibility. After the first jump you make two more tandem skydives, each starting with about 30 minutes of ground training. Those tandem jumps teach you how to make turns in freefall, and you will add body awareness skills and altitude awareness. When you finish the three tandem jumps you will have spent almost 15 minutes of actual flight time learning parachute control with your instructor, who will provide one-on-one training. The tandem part of the training program is designed to build your skills so that you can quickly master the AFF component.

After your third tandem jump many schools add a 6-hour ground class that focuses on advanced flight and emergency procedures. This ground class is similar to the first-jump ground training offered in a standard AFF or static line course, but the material is more in depth, and usually much easier to master because you have already experienced freefall and parachute flight.

Your fourth jump in this program may be an AFF-style skydive, but often with just one instructor in freefall instead of two. This jump is designed to teach you solo flying skills and heading awareness, while exposing you to the fun of flying your own body without a tandem instructor. The fifth and sixth jumps are often release dives, notable because your instructor leaves the airplane holding onto you, but lets go once you have demonstrated body stability. You will probably try some solo freefall turns on these jumps and also try flying your body horizontally through the sky, a technique called "tracking." By the seventh jump you will leave the airplane with an instructor nearby but not holding on. Once in freefall you will work on three-dimensional flight and stability recovery by doing barrel rolls and backloops. The eighth and ninth jumps are often compilation jumps so you can combine all the skills and tricks you have learned through the program. The tenth jump may be a solo skydive, allowing you to leave the airplane without an instructor and just enjoy the freedom of flight with no pressure. The eleventh jump is a solo freefall from about 4000 feet. After that jump you will have a chance to continue skydiving on your own, or you may make more jumps with a coach to build even more skill.

Each jump in the program includes advanced parachute training that teaches you when and where to leave the airplane, and how to navigate your parachute back to the airport.

Photo by Kaz Sheekey

Figure 2.11 Advanced training is designed to allow you to skydive without an instructor.

You will learn about parachute stalls and recovery, spirals and steep turns, slow turns, and how to both sink and float your parachute in the air. You will also learn how to pack a parachute. In fact, you will probably pack your own parachute by about the sixth or seventh jump. Packing often seems difficult and scary to many students, but it is really easy to learn, and once mastered, it will give you added confidence in your equipment.

If you encounter problems with flying your parachute at any point in a hybrid training program, you may have an opportunity to make a static line jump to concentrate on the parachute flight without also worrying about freefall. If the school has a parachute simulator, this could be added to the program too. If your school is near a wind tunnel, you may add some tunnel time to help master freefall skills. Your instructor will be able to use any training methods he or she is rated to teach while building a hybrid training program.

Chapter 3

REGULATION AND INSTRUCTOR CERTIFICATION

I f you are planning to make a parachute jump it is important that you know something about how the government regulates skydiving, and how the industry regulates itself. Although regulation can often seem tedious and boring, a solid understanding of national standards can help you to select a reputable school for your training.

Skydivers tend to be independent people, and as a result, the skydiving community has resisted government regulation. Skydiving instructors do not hold government licenses or certifications, and skydiving schools do not meet any established federal training standards. There are some important laws that govern skydiving, but they are very limited. The skydiving community has developed its own voluntary national guidelines, and most skydiving centers adhere to these suggestions.

FEDERAL AVIATION ADMINISTRATION

The Federal Aviation Administration (FAA) is the only federal agency that regulates the skydiving industry. The FAA is interested primarily in the safety of airplanes and pilots, and in controlling hazards to the general public on the ground. Protecting skydivers and skydiving students is not the primary function of the FAA.

FAA regulations require that commercial skydiving centers use a pilot with either a commercial or airline transport pilot certificate. The FAA requires that a pilot have a minimum of 250 hours of flight time before qualifying for a commercial license, so all pilots flying jumpers should have at least this much flight time. Pilots flying for major airlines typically have many thousands of flight hours. Skydiving pilots are often very new commercial pilots who are working at jump operations as a way of gaining professional pilot experience in the hopes of getting a job at a major airline someday. The skydiving pilots flying at your local school may have as much experience as those flying for major airlines, but the FAA only prescribes the minimum of 250 hours to fly airplanes with jumpers aboard.

Airspace Restrictions

The FAA has always been concerned about skydivers crashing into airplanes in flight, or creating a hazard in the sky, so there are very strong regulations that prevent skydivers from jumping close to clouds, where airplanes might not be easy to see. Jumpers and skydiving pilots are required to look below them and make sure there are no airplanes nearby before jumping begins. Skydiving pilots are also required to talk with air traffic controllers before letting jumpers out of their airplanes, with the hope that controllers will see other planes on their radarscopes.

You may be surprised to learn that the airspace above a skydiving center is generally not restricted to other traffic, and other airplanes and pilots can usually fly overhead without special permission. For that reason the requirement for skydivers to watch for other airplanes and coordinate jumping with air traffic controllers is very important.

Photo by Kaz Sheekey

Figure 3.1   Skydivers rarely jump over clouds.

Parachute Packing

An FAA-certificated parachute rigger is required to inspect and pack your reserve parachute at least every 120 days, even if it has not been used. Reserve parachutes and the accompanying harness system must be certified by the FAA for airworthiness under a program that closely tracks the manufacture of these components. Your main parachute does not need to meet federal manufacturing guidelines. The only major regulation governing main parachutes is that they must be packed by a certificated rigger, a person under the direct supervision of a rigger, or the person using the equipment. It may seem strange, but FAA regulations allow you to pack your own main parachute even though you have never made a single jump, and may have no idea how a parachute works. Fortunately, learning to pack parachutes is relatively easy, and very few people are foolish enough to try packing without at least some training. All schools have staff members who will pack the parachutes you use for your first jump.

The FAA also requires every skydiver to have a parachute system with two parachutes, a main and a reserve, and you can expect every school to comply with this rule.

The FAA began direct oversight of tandem jumping in 2001, when regulations were changed to recognize tandem equipment officially. Prior to 2001 tandem jumping was conducted as an experimental program coordinated between the FAA and equipment manufacturers. More than 2.5 million tandem jumps were conducted under a carefully monitored exemption before the agency agreed to change the regulation.[5] The FAA now requires that tandem instructors have made a minimum of 500 jumps and have been jumping for at least 3 years. Your tandem instructor must be trained and certified by the manufacturer of the equipment being used, the United States Parachute Association (USPA), or by another FAA-approved instructor training method. The FAA also requires that tandem parachute systems have an automatic activation device (AAD) on the reserve, and that the student be briefed about what to do in the event of an emergency.

The FAA does not directly issue licenses for jumping, nor does it certify instructors or skydivers. The FAA relies heavily on the skydiving industry to train and support instructors and other jumpers. Regulations that pertain to skydiving can be found in Part 105 of the Federal Aviation Regulations.

UNITED STATES PARACHUTE ASSOCIATION

Skydivers like the idea of keeping mandatory regulation to a minimum. It makes it easier to develop new programs that enhance safety, and it is always more fun to skydive without being burdened with rules. Many years ago a group of skydivers who were interested in safety got together and formed their own national organization, now called the United States Parachute Association (USPA). The USPA works

Photo by Kaz Sheekey

Figure 3.2   Tandem instructors must meet FAA minimum training standards.

closely with the FAA to ensure the safety of skydivers and students around the country. The USPA is a not-for-profit organization representing more than 34,000 jumpers and about 300 skydiving centers. Almost every active U.S. skydiver is a member of the USPA, and about 80 percent of skydiving schools belong to the group.[6] The USPA publishes a short list of mandatory Basic Safety Requirements (BSRs), a series of comprehensive instruction manuals, and a monthly magazine distributed to all members. (See Figure 3.3.) The USPA also issues licenses and special ratings to skydivers and instructors, and publishes guidelines for advanced skydiving. The organization has even arranged for third-party insurance that covers the cost of accidental damages caused by USPA members. This insurance, provided free to all members, is designed to pay for property damage in the event of a parachuting mishap, such as a skydiver landing on a car that is owned by somebody else.

Figure 3.3   The *Skydivers Information Manual* is an important publication distributed by the USPA. (© *USPA.*)

It is important to understand that USPA is a voluntary membership organization. Not all jumpers are members, and not all skydiving schools follow USPA requirements. Those schools that do belong to the organization have agreed to follow all the BSRs and to adhere to other specific standards.

Most skydiving centers have agreed to recognize USPA-approved skydiving training conducted at another school, and to recognize licenses issued by the USPA. Once you start your training it is a good idea to stick with one program until you have obtained a skydiving license, and then you will be welcome at other drop zones by showing proof of that license along with a USPA membership card. Most jumpers buy their own equipment when they qualify for a license, and that makes traveling to other drop zones a whole lot easier. The first license available to USPA members is called an "A" license, and it can be earned after about 25 jumps.[7] When you have more experience you may qualify for "B" and "C" licenses. The highest level license available is the "D" license, earned after about 500 jumps.[8]

Basic Safety Requirements

BSRs mandate that all skydivers complete a standard form stating they are in generally good health. Most skydiving centers build this into a liability waiver that every visiting skydiver and student signs. BSRs also mandate a minimum age for skydiving. Many skydiving centers have established their own minimum age, usually 18.

BSRs also list minimum instructional requirements for student jumps in the three basic types of training programs (tandem, Accelerated Freefall, and static line), and offer brief guidelines defining requirements for crossover training when changing between programs. The BSRs mandate that schools use instructors that have earned USPA ratings to teach in the specific program being offered, and that the instructors follow established guidelines. There is an exception made that allows non-USPA-rated instructors to take students on introductory tandem jumps, but they must be certified by a tandem equipment manufacturer or another method approved by the FAA.

The Basic Safety Requirements mandate that all student skydivers be equipped with an altimeter, an automatic activation device (AAD), a reserve static line (RSL), a square-style main parachute, and a steerable reserve parachute. The BSRs also require that every student have a rigid helmet, but this requirement is waived for tandem students.

Wind limitations, opening altitudes, and drop-zone obstructions are also covered in the BSRs, with specific restrictions based on experience level. Solo students are limited to winds of 14 mph, or 10 mph if their reserve is the round style. Tandem students are not subject to wind limitations because they fly with an experienced instructor. Tandem students must open their parachutes no lower than 4500 feet above the ground, and solo students must open by 3000 feet.

In addition to the BSRs, the USPA publishes an outline for its Integrated Student Program (ISP). This program defines a high-level standard for student training, and establishes a jump-by-jump framework for meeting that standard. The ISP is designed to allow you to easily move from one training method to another, with specific training objectives that can be met on tandem, static line, and Accelerated Freefall jumps. The ISP is designed so you can complete all your training in one method, or build a hybrid program with different types of jumps. The ISP is not a mandatory program, but many skydiving centers use it, while many others combine parts of the ISP with their own program. When you talk with skydiving instructors they should be able to tell you how their local program compares with this national standard.

USPA Ratings

The USPA also issues instructor ratings in several programs. Some of these instructor ratings are easy to earn, and others are very difficult to obtain.[9]

The first level of USPA teacher is called a "Coach." This position is available to skydivers who have made at least 100 jumps and who have completed a special training program. Coaches are authorized to assist with ground training and to fly with advanced students, but they cannot jump with first-timers. Coaches are often very enthusiastic teachers and generally work closely with rated instructors.

The USPA "Instructor" rating is issued for a specific type of program and requires far more training and experience than a Coach rating. Ratings issued cover tandem, static line, instructor-assisted deployment, and Accelerated Freefall. Many instructors have earned ratings in several different programs. Your instructor will probably be happy to tell you about the training he or she has received.

A static line or instructor-assisted deployment (IAD) instructor must have made at least 200 jumps. New static line and IAD instructors must begin their training with a Coach course, and then take another course that is specific to the static line or IAD method.

A USPA "Tandem Instructor" rating is more difficult to obtain. It requires a Coach rating, at least 500 jumps, a training class in tandem teaching, a training class that covers the specific model of tandem equipment being used, and a series of jumps as part of an apprentice program. Most new tandem instructors receive their training at a tandem certification course offered by a USPA Tandem Course Conductor who is also rated by the manufacturer. Basic Safety Requirements dictate that Tandem Instructors who have been rated by the USPA must also hold a medical certificate issued by an FAA-certificated physician.

The most difficult USPA instructor rating to earn is the Accelerated Freefall rating. The AFF instructor rating requires at least 6 hours of logged freefall time, a Coach rating, and completion of a very demanding certification course.

The USPA also trains experienced skydiving instructors to run special courses for other skydivers who want to become instructors. Course Directors, Instructor-Examiners, and Evaluators have completed even more training than regular instructors. You will find that most instructors are very proud of their ratings. They will be happy to talk with you about their certifications and what training they have received.

A Safety and Training Advisor (S&TA) is a volunteer appointed by the USPA Regional Director to cover a specific drop zone or area, and is responsible for the administration of USPA programs within that area. An S&TA is generally a

Figure 3.4 A USPA Safety and Training Advisor is responsible for administering USPA programs at a specific drop zone. (© *USPA*.)

senior skydiver with solid knowledge of FAA and USPA rules, as well as a wide variety of skydiving experiences. An S&TA is sometimes the drop-zone owner, and sometimes just a regular jumper with an interest in student training. If you have questions about the programs at a USPA drop zone, the S&TA will probably be able to provide you with well-considered answers.

SKYDIVE UNIVERSITY

USPA instructional programs are a voluntary minimum standard for all skydiving schools in the country. A higher standard has been created by Skydive University, a for-profit company that develops proprietary skydiving training programs and issues its own Coach ratings.

The Skydive University program uses a well-developed syllabus, one-on-one mentoring, classroom training, ground practice, muscle memory, and visualization to help you develop advanced skills. Several drop zones around the country have been designated as Skydive University campuses. These campuses offer a complete training program to take you from your first jump through the development of advanced flying skills.

Some Skydive University Coaches work independently of the campus system, and can be found teaching on other drop zones around the world. If your instructor has a Skydive University rating you can be sure it required lots of extra training, as well as the demonstration of advanced air skills and teaching ability.

In addition to advancements in conventional training, Skydive University has developed a unique wind tunnel flight training syllabus that substitutes tunnel time for actual skydives. Using this program will allow you to complete the basic USPA Accelerated Freefall requirements in fewer jumps, and make your very first solo with just five real skydives.

The Skydive University Coach rating is more difficult to earn than the USPA Coach rating. A Skydive University Coach must have a minimum of 200 jumps, complete a 4-day training course, pass a skydiving skills test, a teaching evaluation, and a video analysis skills test. Following the course and testing, a jumper works as an apprentice Coach for at least 50 jumps. After the apprentice period the jumper applies for final certification as an official Coach.

STATE REGULATION

A few states have established regulations covering skydiving, but the regulations are usually very limited. Some states, such as New Jersey and Connecticut, require drop zones and skydiving schools to register with the state, and apply for an annual skydiving permit. The skydiving permits are generally simple to obtain and rarely place any serious restrictions on the school. Nevada has established an interesting regulation that requires all drop zones within the state to comply with USPA Basic Safety Requirements, and to report all serious injuries to the USPA. Several other states have their own regulations, but the regulations are generally very limited, and it will frequently be difficult for you to determine what they are. The best way to find out about state regulation is to ask the manager of the sky-diving school what laws govern its operation.

TANDEM MANUFACTURERS

There are currently three manufacturers producing tandem parachute systems for use in the United States, and a fourth that has stopped selling new equipment. Each of these man-ufacturers has received approval from the FAA to train and certify instructors to use their own models of tandem equip-ment. An instructor certified by one manufacturer is only allowed to jump tandem rigs made by that manufacturer, unless he receives additional certification to use other tan-dem rigs. There are some significant differences in the way tandem gear is designed, so it is important that an instructor be trained in the specific nuances of the rig being used. Your tandem instructor may have received training from just one of the manufacturers, or may be rated to use several tandem systems. In 2002 the FAA approved a program that allows the USPA to train and certify tandem instructors without requiring the instructor to have additional certification from the equipment manufacturers. Instructors who have been trained under the USPA program are also rated for specific tandem equipment, and must receive additional training if they want to jump with tandem equipment made by another manufacturer.

Because of recent regulatory changes, tandem instructors have many different levels and types of certification. It is possible that your instructor has been certified by the USPA but not the manufacturer, or he may have been certified by the manufacturer and not the USPA. Some instructors may even hold ratings from the USPA and several manufacturers. Your instructor should be able to tell you what specific certifications he has earned, and what rules govern the equipment you will be jumping.

Some tandem equipment manufacturers maintain tight control over how their rigs are used and require instructors to follow very specific rules. In a few cases these rules have been specified in writing as part of a purchase agreement when the tandem rig was sold to the school. Some of these manufacturer rules pertain to minimum age limits, maximum weight, equipment maintenance, and compliance with USPA BSRs. Not all

Figure 3.5   Your instructor will be able to tell you about his training.

tandem manufacturers have established specific policies or mandatory rules. A local skydiving school may be bound by the tandem manufacturers' requirements, or may not follow any manufacturer guidelines at all. You should feel comfortable asking the school about what manufacturer guidelines cover the equipment you will be jumping.

Chapter 4

PSYCHOLOGY

Jumping out of an airplane for the first time can be very frightening. Every student will experience some level of fear, and a few students will be so terrified that they are unable to enjoy the jump at all. Most likely your fear and anxiety will swing through peaks and valleys as you prepare to make your first skydive. It might help you to know that stress is normal, and can be controlled to some degree.

YOUR STRESS LEVEL

If you have never participated in a high-risk sport you may find skydiving especially stressful, but you will be able to develop emotional skills to help manage the stress, and those new skills can often help you in other parts of your life. Students who have participated in high-risk physical activities such as rock climbing, motorcycle racing, or scuba diving tend to have a much easier time adjusting to the anxiety of skydiving.

Many people would like to make a skydive, but they don't think they will have the courage to actually jump, so they avoid the sport and never make an effort to skydive.

Often people who avoid one activity, such as trying a skydive, also avoid other things in life that could bring them enjoyment. If you can get over the hurdles and try just one activity that you find terrifying, it will often help you to develop the strength to conquer other fears too. That is one reason why people who participate in activities such as rock climbing have an easier time adjusting to skydiving. They have already mastered one fear, so they know they have the ability to conquer another.

Believe it or not, some experienced skydivers were once afraid of heights, and a few still are. That probably sounds strange because skydivers jump from so high up, but frequently the extreme distance above the ground reduces depth perception, making the visuals easier to deal with, and that lessens the fear. People who successfully jump from an airplane often find that once they have made a skydive, their fear of heights diminishes and becomes easier to handle. The change doesn't usually happen overnight, but rather it follows consistent efforts at dealing with the specific fear. Skydiving can easily be a part of that effort. Frequently, a person who deals with one fear, such as fear of heights, has a much easier time dealing with other fears, such as public speaking, or just about anything else that causes stress. You may find that a skydive will help you discover your own inner strength, and give you the confidence to try other exciting new activities.

A MODEL OF FEAR MANAGEMENT

One of the most challenging parts of dealing with the stress of a skydive is figuring out what is causing the stress in the first place. Skydiving is certainly a risky sport, and many people are afraid that something might go wrong, but they rarely try to identify their fear beyond that broad definition. If you are feeling apprehensive about skydiving, or any other high-risk activ-

Figure 4.1 Skydiving can improve your self confidence and help you enjoy life to its fullest.

ity, it may help to break down the activity or event into isolated variables.

You can easily model an activity such as skydiving in terms of your own skills and responsibilities, other people's skills and responsibilities, the equipment you will be using, and the environment. Once you have broken an activity into those four categories, you should make an effort to learn everything you can about each one, and how they relate to a successful skydive. (See Figure 4.2.)

Self

The first element of the model is *self*. When you begin thinking about your role in the skydive, you should consider what is really going to be required of you, and if you have the necessary

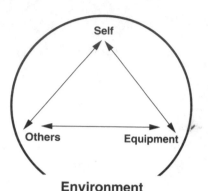

Environment

Figure 4.2 Isolating variables will make it easier to deal with fear.

skill to complete the anticipated tasks. In the case of a tandem skydive, you really don't have to do much of anything. Heck, even if you don't do anything at all, a good instructor will be capable of dealing with the jump alone. You may be given some learning tasks, or be asked to help out on the jump, but your responsibilities are very limited, so you should not be afraid of personal failure on a tandem skydive.

If your jump is going to be a static line or Accelerated Freefall jump, you will be responsible for your own safety, and will need a great deal of knowledge and training, and with that will come some very specific practical skills. If you have already participated in similar sports, or made a tandem skydive, you will be in a good position to evaluate your own aptitude for solo skydiving. If you know you have the basic aptitude for the sport, it becomes a simple matter of obtaining the needed training. Once you have the training, you will be able to self-assess your skills, and if you think you are not ready to jump, you can always ask for more training.

The ground course in an Accelerated Freefall or static line program will always include a practical test or demonstration of skill, usually near the end of the training. The purpose of that test is to show the instructor that you have mastered the material and are ready to skydive. You should also take advantage of that testing block to evaluate your own skill. If you know you easily handled the testing at the conclusion of your training, you should have confidence in your skill, and that self-confidence will help reduce stress. If you are struggling with the demonstration of skill, you should seek additional training until you are personally satisfied in your own abilities. One key to controlling stress is to be self-confident, and that requires that you meet your own standards of performance.

If you know what is expected of you on the skydive, have identified critical skills needed for success, and have received appropriate training, you should have confidence that you can meet the performance standard. If you know what the standard is, and know that you can meet the standard, you will have an easier time controlling stress on your skydive.

Others

Skydiving is not a solo sport. Other people are always involved, especially as you are learning. Your ability to deal with problems in the air will be dependent to a very large degree on the skill and experience of your instructors. If you trust your instructors, you will find it much easier to relax and enjoy the skydive.

A good instructor will take a moment for introductions, and should provide you with a quick overview of his or her experience and training. If your instructor does not tell you about himself or herself, you should ask very direct questions so that you know he or she is qualified to handle the requirements of the skydive. Your instructor should be knowledgeable, and should show a genuine interest in your questions.

If you are making a tandem skydive, you should gain confidence that your instructor can handle the flying tasks, and deal with any emergency. If you are not comfortable with your instructor, you should think about what will give you the

Figure 4.3   You should feel comfortable with your instructor.

needed confidence, and then find a way to build your comfort level so you can relax. Sometimes that can be as simple as speaking with another staff member at the drop zone. You may learn that your instructor is well respected, experienced, and is a very good skydiver with great skills. Sometimes it helps to hear that kind of assessment from a second source.

If you really don't trust your instructor or don't have confidence in his ability, you will probably be unable to relax and enjoy the skydive. In that case, you should ask the school for another instructor. It is sometimes difficult to speak up about your concerns, but as a tandem student, your key objective should be to relax, and that requires that you have confidence in your instructor. Most schools will try to accommodate your requests, either by changing instructors, or helping you to relate to the instructor who has been assigned. When you make a tandem skydive, it is critical that you trust your instructor, so do whatever it takes to build that trust.

If you are making an Accelerated Freefall or static line jump, you will need to have confidence in your instructor's ability to communicate and teach, as much as in his or her skydiving skill. Ultimately, you will be responsible for managing the jump yourself, but the instructor will need to give you the skills necessary to do that. When you finish the training course you should be comfortable that you have been given all the information and training you need to handle emergencies on your own, and that your instructor is capable of assessing your skill adequately.

As with a tandem skydive, you should expect AFF and static line instructors to introduce themselves, and you should be comfortable that they have the required experience and training. Instructors in the AFF or static line program need to have lots of skydiving experience, but more than that, they need to be able to relate to you as a student. Much of your success will depend on how well you have mastered the material, and that will frequently be dependent on the relationship you have established with your instructors.

Obviously, you should do whatever it takes to develop a good working relationship with your skydiving instructors. Unfortunately, we occasionally allow our own biases to interfere with building the trust needed to learn. Often we seek teachers who are similar to us, and can have a hard time receiving instruction from people who are different. Some male students have a hard time learning from women instructors, and some female students are uncomfortable with men as instructors. It is important for you to know that most skydiving instructors are equally qualified to teach, and that their sex or personal background should have little bearing on the instructional relationship. You should endeavor to build a solid learning relationship with your instructors based on their teaching style and knowledge. You may find that instructors who are different from you can actually explain material in new ways to which you might not otherwise be exposed.

Students at a small drop zone may complete an entire 20-jump training program and never have the opportunity to jump with more than one instructor. That can be really tough if you don't relate well. Other students have many different instructors through their training, and they usually find one or two they like better than the rest. You should feel free to ask for specific instructors as you gain experience, but also be open to the benefits of the different presentation styles offered by the various instructors at your drop zone.

If you have a good instructor and make a little bit of effort, you will build a great relationship, and that will give you confidence as you prepare for your skydives.

Equipment

Skydiving is an equipment-intensive sport, and you will want to be sure the parachute gear you are using is in good condition. It helps to know something about the gear, and your instructors should be able to show you quickly how it works.

Figure 4.4   Try to build trust with your instructor. Many drop zones have both male and female instructors.

You should feel free to ask questions so you can gain confidence in the equipment. If you are making a tandem skydive, it may be enough for you to know that your instructor has an established set of emergency procedures, and your reserve parachute has a computer attached that is programmed to open it in the event of certain problems. You may also want to know how old the gear is, who packed it, and why the school selected that model of equipment over competing brands.

If you are making a static line or AFF jump, you will need much more detail about the equipment in order to feel comfortable making a jump. You will obviously need to know what to do in an emergency, but it also helps to know how the equipment works, and why pulling specific handles in a specific way is important. You will also want to learn about how old the gear is, how it is maintained, and why your instructors have confidence in the equipment they are providing.

Environment

Finally, you will need to be comfortable in the environment. It is important for you to understand that some weather conditions may not be appropriate for skydiving. High winds or a low and solid bank of clouds make jumping dangerous, and should be avoided, although some clouds, and a bit of wind, will be no problem at all. In fact, a light breeze will improve the landing performance of most parachutes, and that is especially true for tandem parachutes. If you are jumping on a very windy or gusty day, you may find yourself uncomfortable with the conditions, and that can create stress. If you think environmental factors may be a problem, you should seek guidance from your instructor, or other members of the school staff.

As a guide, AFF and static line students generally should not jump in winds greater than 14 mph. Tandem students can jump in higher winds because they are flying a bigger parachute with an experienced instructor. Most schools will not take students on tandem skydives if the wind is much above 25 mph, but that decision will depend on other factors too, including turbulence. Many schools prefer to have a light breeze to make tandem landings safer and more comfortable. If you have questions about the suitability of weather conditions, you should ask your instructor.

Using the Model

Your stress level will probably be quite high as you begin to think about making a skydive, and you start contacting schools. You can reduce that initial stress by learning as much as possible about what will be required of you, the qualifications of skydiving instructors, the equipment that will be used, and the appropriate weather conditions for skydiving.

Once you arrive at the school your stress level will probably peak again, and that is a great time to further build your knowledge and gain trust in the specific instructors who will be working with you. Once again, do a quick inventory of your own skills and aptitude, the qualifications of the other people involved, the quality of the equipment, and the environmental conditions of the day. If you have questions or concerns, get them addressed to your satisfaction. The extra knowledge you gain at this point will help to reduce your stress level now, as well as keep your stress in check as you get ready to jump.

Most students experience another peak of fear just as they are getting ready to jump out of the plane, and continuing for the first few seconds after the exit. As you approach this phase of the jump you should be well armed with knowledge and trust. As the stress builds, take another quick inventory of self, others, equipment, and environment. After you have analyzed the jump using that model, you will probably be able to relax a bit and then just go with the flow. If you are making a tandem skydive, focus on the "others" part of the model. If you are making a static line or Accelerated Freefall jump, focus more heavily on the "self" part of the model. Take a moment on the plane ride to ask yourself: Am I comfortable with the training I have received? Am I comfortable with the skill and experience of my instructor? Am I comfortable with the equipment provided? Am I comfortable with the weather conditions today?

Almost every student will feel some stress as the actual jump gets closer. If you have analyzed the key variables, you may find that fear easier to control. If there are elements that you are uncomfortable with, such as an inattentive or seemingly unqualified instructor, equipment that looks to be in poor condition, or strong winds, you should not make the skydive. You are ultimately responsible for your own safety and should obtain enough knowledge and experience to decide if skydiving is a reasonable activity for you to participate in. If you decide that skydiving is not appropriate for you, or that you are not happy with your training, the support staff, the equipment being used, or the weather, you have a responsibility to remain on the ground or in the airplane, and not jump. Making that decision is difficult, but with the use of the model you should have an easier time isolating variables and defining where your stress is coming from. That should give you a reasonable way to correct whatever the issues may be, and gain the needed confidence.

Almost any activity can be analyzed with the model of self, others, equipment, and environment. It is a useful way of understanding and controlling the variables of a complex physical activity, and analyzing the source of your own fear. If you are successful in using the model to gauge the appropriateness of skydiving, you may find it useful when participating in other activities too.

USING FRIENDS TO CONTROL STRESS

In spite of your best efforts, there will certainly be some stress as you jump from an airplane for the first time. There are a few other ways that some people have found to help control their fear. Most students arrive at the drop zone in groups and use friends for support as the stress builds. Some people like to be the first in the group to jump so that they have the added pressure of their friends jumping next. Others like being at the end of the group so they can watch a few people first and gain extra confidence before jumping, and have less peer pressure when it is time to leave. If you are making your skydive with friends, it is a good idea to talk with them in advance about jump order, and use that as a mechanism for controlling your own fear.

Some people are more comfortable if they arrive at the drop zone and jump all alone, with no friends along as distractions. If you are generally independent, you may be less self-conscious and more comfortable making your first jump alone.

For some people it helps to have a camera person recording the event so that you feel an obligation to perform, while other students don't want the distraction of a camera, or may feel it will make them overly self-conscious.

RELAXING

It always helps if you have something on which to focus as the jump gets underway. That will keep your mind busy with directed thoughts, rather than letting it wander and focus on the natural fear. You will probably find that the airplane ride will be a bit slow, and you will have time to think through the jump procedures. Some students like to imagine everything about the jump in great detail, and visualize exactly what they think it will be like. If you can imagine the event accurately in advance, your mind will be better able to handle the experience when you actually jump. Accurate visualization will be enhanced if you have already seen video of a skydive, or if you have great descriptions of what will happen and what it will feel like from a beginner's perspective. With a solid background of realistic visuals, it only takes a good imagination and a moment of quiet to pre-create the jump experience in your own mind.

Some students find it helpful to close their eyes while riding in the airplane and think about something pleasant, perhaps a quiet place in the woods, or a recent evening with friends. Try to think about the detail of that place or event as a way to take your mind off the stress of the skydive. The more detail you can visualize, the easier it will be to control the stress. If possible, you should try to smile while thinking of the pleasant activity. Smiling can have a soothing effect on your nerves all by itself.

Figure 4.5   Try to relax in the airplane.

Once the airplane is at the jump altitude and approaching the airport, it will begin to feel like things are moving much faster. Your instructor will be using the quickened pace to help you focus on the jump, and to eliminate distractions. At that point, listen to your instructor, take a couple of deep breaths, relax your muscles as much as possible, smile, and think about how much fun you are about to have.

Most students find that their fear level peaks as they approach the door, then subsides quickly once they are in freefall. If you find your stress level increasing as you prepare for the skydive, try to relax and anticipate the enjoyment of freefall.

USING CALL-OUTS

If you are making a tandem skydive, your instructor will probably have a very specific set of things he would like you to do as you leave the airplane, almost certainly starting with a good body position called an "arch." Your instructor may even shout out "Arch" as you leave the plane. If you join him in saying the word, even if it is just saying it to yourself, you will be more likely to arch. The idea is to use your voice and word selection to drive a specific action or behavior. Many instructors use a series of exit commands such as "ready, set, ARCH" to get the exit going, and help your brain to focus on the required tasks. Saying the word "arch" out loud is much more likely to drive an appropriate physical response than screaming something like "Geronimo," an irrelevant word that you may have heard in some old Hollywood skydiving films.

If you are making a static line jump your instructor will probably give you a series of shout-out exit commands such "ARCH, 1,000, 2,000, 3,000, 4,000, Check," as a way to focus your mind on the arch first, then build a sense of time awareness, and finally drive you to check your parachute visually. As with the single command of "Arch" used by many tandem instructors, the more complete exit call-out helps your mind to focus.

Accelerated Freefall students will probably be given even more shout-outs to use through the skydive, including a verbal call-out of altitude when reading an altimeter. You will find that your awareness will increase if you articulate a specific behavior or action. If you shout the word "Arch," you will be more likely to actually arch. If you shout the word "Check," you will be more likely to actually check your parachute. When you read your altimeter and shout out whatever it says, you will be more likely to understand and respond to the actual altitude.

Think for a moment about the last time you checked your wristwatch, then an instant later somebody asked you what time it was. Most likely, you had to look again, because the actual time didn't really register. If you had told somebody what time it was when you first read your watch, your brain would probably remember that detail. Skydivers frequently use this knowledge of verbal call-outs to maintain their awareness and improve recall of the skydive. It is not uncommon for a skydiving instructor to talk through what he or she is doing and seeing in freefall, often out loud. You won't be able to hear your instructor's voice, but by simply talking to himself or herself, the instructor will use the key words to direct his or her brain and actions, and improve awareness and recall. You can also use this skill by talking to yourself throughout the jump. Some students like to call out whatever they are looking at or thinking about, while others like to build a silent narrative that will help them remember the whole jump later in the day.

CONVERTING STRESS

For many people, leaving the airplane may feel very stressful, or scary, but often that stress is actually a powerful and positive form of excitement. If you think of an actor about to begin a performance, you may be able to imagine the stage fright that he or she might feel. In many cases, an actor can turn that fright into excitement, and use that new excitement to drive his or her own performance. You can do the same thing with your emotions when you are about to make a skydive.

Sometimes the difference between fear and excitement can be as simple as the visuals you are thinking of, or the outcome you expect. Many students can convert their fear to excitement by changing the pictures in their minds from negatives to positives. If you find yourself worried about the threat of an accident, try changing your expectations to something more positive, such as the fun you will have, or the sensation of flying. You might even think about how you will brag to your friends after the jump, and what kind of stories you will tell them.

The key to success is to develop control over your own mind, and shape the way you deal with your emotions. That can be as simple as reframing your expectations or changing the way you previsualize the jump. If you expect to "freak out," you very well might. However, if you expect to have fun, you will probably find it easier to relax, and will have a much better skydive. As you visualize the jump, think about the fun things you will be doing. Imagine the clouds moving by, the relaxing sensation of freefall, or the sweet smell of the air thousands of feet above the ground. Imagine yourself with a smile on your face as you take in the experience of a lifetime.

IMPROVING YOUR LIFE

Perhaps the best part of skydiving is that it can empower you to seize control of your emotions. Many people initially believe they will not have the courage to make a jump, but by using appropriate stress management techniques they are able to enjoy skydiving. It is important to understand that fear, stress, and anxiety are emotions that are created in your own mind, and you have the power to control them. There is an expression that says "there are no victims, only volunteers," and it certainly applies to skydiving. You can let your emotions control you, or you can control your emotions.

You probably already have some ways of dealing with fear and anxiety, but you will develop others as you make that first jump. Many students, especially in the Accelerated Freefall and static line programs, gain tremendous self-confidence from skydiving, and often they develop new emotional skills that can be applied to other aspects of their lives. The key to success in skydiving is to gain the knowledge you will need to relax, then take control of your own emotions, and enjoy the unique experience of human flight.

Photo courtesy of Archway Skydiving

Figure 4.6 You have the power to control your own emotions.

Chapter 5

EQUIPMENT

When you make your first jump the skydiving school will provide all the equipment you need. Most skydiving schools will continue to supply the required equipment throughout your training program. Once you have completed training you may have an opportunity to rent most of the gear, or you may be required to buy your own. The equipment used by students is slightly different from the gear used by experts, but the core functions are the same. Although it is not critical that you know very much about skydiving equipment when you make your first jump, some students find it comforting to understand their gear in great detail. Other students are interested in the unique attributes of the gear they see being used by more experienced jumpers around the drop zone. Some beginning students may also be looking ahead to the day when they are jumping on their own, and may think about buying equipment before they even make their first jump.

A COMPLETE RIG

The most important piece of equipment is the rig. This is the backpack-style system that holds the parachutes. A complete rig is actually composed of a harness/container system, a main parachute, and a reserve parachute, and each is usually designed and sold as a separate component. A rig may also include an automatic activation device (AAD) that is designed to deploy the reserve parachute in an emergency. (See Figures 5.1 and 5.2.) The FAA requires that all parachute systems used for intentional jumping have two parachutes, a main and a reserve. The harness/container system and the reserve parachute must be approved by the FAA for use in the United States. Many other countries have their own approval process for this equipment.

In addition to a complete rig, a properly equipped jumper will need an altimeter, jumpsuit, goggles, helmet, and a log book. Some jumpers also carry an audible altimeter that beeps at a preselected altitude as an extra warning when it is time to open their parachute.

Figure 5.1   A complete rig (front).

Figure 5.2   A complete rig (back).

Skydiving equipment is readily available from special-ized dealers. Many dealerships maintain retail stores at large skydiving centers and have mail-order catalogs or web sites. Large dealerships generally stock everything you will need, including complete parachute rigs. Some dealers at larger drop zones have programs that allow you to try different equipment before you buy it, or rent for a while and apply the rental fees to the purchase of your own gear. Large dealerships also carry a wide assortment of skydiving videos, magazines, and T-shirts that will appeal to first-time jumpers. Most small drop zones have at least one or two jumpers who act as dealer representa-tives, and they can help you obtain any needed equipment. Smaller dealers may have a more limited supply of equipment, and fewer rental or demo options.

TANDEM EQUIPMENT

A tandem rig always has two very large parachutes that have been designed specially for the unique needs of tandem jumping.

A tandem system also includes a separate harness for the student that is manufactured by the same company as the harness/container. The FAA requires that all tandem reserves be equipped with an automatic activation device that is properly maintained, and that has been approved by the manufacturer of the tandem equipment.

The tandem harness is a full-body harness. It wraps around your shoulders and thighs, and has straps that go across your chest. Your instructor will adjust the harness while you are standing on the ground so that it is very snug, with just a bit of play in the chest-strap area. It may feel loose when you are sitting in the airplane, but it will feel tighter when you are in the air. The harness will have an attachment on each of your shoulders, and one on each hip. (See Figure 5.4.) The tandem harness is designed to spread the load of the parachute opening over the four separate attachment points. Each of the attachment points is designed so that in an emergency just one

Figure 5.3   Your harness will be adjusted to fit.

can be used alone, but your instructor will use all four connections on every jump. You may find it comforting to know the harness is so well constructed that just one attachment point is enough to keep you connected to your instructor.

The harness your instructor wears has a container on the back that holds the main and reserve parachutes. The container is designed as two separate parts, with the reserve parachute enclosed in the top section, and the main parachute in the bottom. The main parachute is rectangular in shape and has about 350 to 500 square feet of surface area. It is more than twice the size of a typical parachute used by a single experienced jumper. The reserve parachute is about the same size as the main.

Throwing the Drogue

Shortly after you leave the airplane the tandem instructor will probably use a drogue chute to control the speed of your freefall. The drogue is a small round parachute that is attached to your instructor's rig with a long bridle line, and then to the pin that holds your parachute closed. (See Figure 5.5.) The

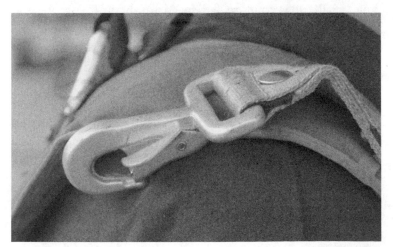

Figure 5.4 A tandem harness has four attachment points.

Photo by Craig O'Brien

Figure 5.5   Your instructor will use a drogue to control freefall speed.

drogue will slow your freefall speed to about 120 mph, and when you pull the ripcord it will help the parachute open. The normal freefall speed of a single jumper is about 120 mph, but two people jumping together without a drogue will descend at about 180 mph. Such high speed is stressful on the equipment and can make the parachute opening very uncomfortable, so it is important for the instructor to use the drogue. Some instructors may be willing to hold off on throwing the drogue for a few extra seconds so you can feel what high-speed freefall is like, but many schools prefer not to do this on a first jump.

When the tandem ripcord is pulled, it releases the pin that holds the drogue to the rig attachment point. The drogue then acts as a giant anchor, and as the bridle line is extended it pulls the parachute off your instructor's back. Tandem parachutes are designed to open over a few seconds, helping to dissipate your freefall energy over a short period of time. The openings are not usually very hard or uncomfortable. Once the main parachute has opened, the drogue trails behind your parachute. When you are at the drop zone watching other tandem parachutes land, you may see a drogue trailing limply behind the inflated main parachute.

Tandem parachutes are designed to land softly, but occasionally landings can be a bit uncomfortable. The quality of the landing will be governed by the type and age of the parachute, the combined weight of you and your instructor, the weather conditions, and the skill of your instructor. If you are helping your instructor by flaring the parachute with him, your efforts will also affect the landing.

When you look at a tandem parachute rig you will see lots of handles and controls. The instructor's harness has two ripcords for the main parachute so that if one ripcord is not available or is not working, the other can be used. It also has a handle on the right side that is attached to a mechanical connection holding the main parachute on the harness. If the main parachute does not work properly, your instructor will pull that handle, and it will release the main parachute from his harness and send you back into freefall. There is another ripcord on the left side of the harness for a reserve parachute. The reserve parachute is designed and packed so that it will open very quickly when needed. All tandem rigs also have a reserve static line (RSL), an emergency device that connects your main parachute to the reserve ripcord so that when your instructor releases the main, it automatically pulls the reserve. On some rigs the RSL can also be pulled by your instructor as an alternate reserve ripcord, if the regular reserve ripcord is not available. The extra handles and emergency devices can make a tandem parachute system much more difficult to operate than a regular rig, but they also give your instructor more emergency options, and may improve safety.

Manufacturers

Just three companies make tandem equipment approved for use in the United States. Strong Enterprises and The Uninsured Relative Workshop developed tandem skydiving in the early 1980s and remain the largest suppliers of tandem equipment. Strong Enterprises makes a rig called the Dual Hawk, and The Uninsured Relative Workshop makes two rigs, called the Vector and the Sigma. The Sigma is the newest tandem rig on the market and features some impressive safety and comfort features. The Jump Shack makes a rig known as the Racer, which has a very comfortable student harness and a few unique design elements that some instructors really like. A fourth manufacturer called Stunts Adventure Equipment was producing a rig called the Eclipse that is very similar to the Vector, but they have stopped production and no longer sell new equipment. A few skydiving schools still use the Eclipse system. Your instructor will be able to tell you what kind of tandem rig you are jumping, and he will almost certainly be able to explain some unique advantages of that system. Skydiving instructors know that some students develop extra confidence when they understand the parachute gear they are jumping, so you should feel comfortable asking about it.

A complete tandem rig will weigh between 40 and 50 pounds and cost between $8000 and $11,000. That is a very large capital expense for most skydiving centers. Small skydiving schools usually have just one or two tandem rigs, but large schools may have several dozen.

STUDENT PARACHUTE RIGS

Parachute rigs used by solo students for static line or Accelerated Freefall jumps are very similar in appearance to tandem equipment, but much smaller. The main parachute is rectangular in shape and has roughly 200 to 300 square feet of surface area. The reserve parachute is roughly the same size as the main you are using. The skydiving school will select parachutes that are well matched to your body weight, with heavier students using larger parachutes. Reserve parachutes are generally also rectangular, but a few schools use round reserves. If the school where you are training uses round reserves, it should provide you with extra instruction on how to fly that type of parachute, and will limit you to jumping in light wind conditions.

Student rigs do not have drogues, but they do have a similar device called a pilot chute. A pilot chute is used only to open the main parachute, and not to control speed in freefall. A pilot chute is a small round parachute that measures about 25 to 36 inches in diameter. Some schools use a hand-deployed pilot chute located in a pocket on the bottom of the rig, while others use a spring-loaded pilot chute that is packed inside the main container. When it is time to open the main parachute, you either pull the hand-deploy pilot chute out of the pocket and toss it into the air stream, or you pull an old-style ripcord, and the spring-loaded pilot chute shoots off your back and into the air. Once the pilot chute fills with air, it acts as an anchor, and an extending bridle instantly pulls the main parachute into the sky where it inflates. Most parachutes are actually packed inside a small bag, with the lines held to the bag by rubber bands. As the bag is pulled from the harness/container system, the lines extend, and then the parachute is exposed. Once the parachute material is in the high-speed air, it inflates. After your parachute opens, the pilot chute and bag simply trail behind it. (See Figures 5.6 and 5.7.)

Square Parachutes

A rectangular parachute is also called a square parachute, or a ram air parachute. It is made of nylon and composed of a series of cells that are open in the front and sewn shut in the back. As

Figure 5.6 It is easy to use a hand-deployed pilot chute.

Figure 5.7 The pilot chute pulls the bag from your container and starts the deployment process.

the parachute fills with air, it takes shape and slows your rate of descent. The shape is maintained by air being forced into the cells in the front, and by lines of various length that attach the parachute to nylon risers, and then to your harness. The parachute behaves like an airplane wing by creating lift as it moves through the air. Steering the parachute is accomplished by pulling down on toggles on the lines attached to the back of the parachute. When you pull on the left toggle it pulls the left side down and makes the parachute turn left. When you pull on the right toggle it makes the parachute go right. When both toggles are pulled together the parachute will actually pitch up slightly, and the extra aerodynamic lift will reduce forward speed and vertical descent, sometimes even allowing the parachute to go back up by just a few feet. Pulling both toggles together is called flaring, and is the best way to land a square parachute. Your instructors will have you practice this on the ground and in the air so you can use the technique when you land.

Other Components

A student rig generally has a cutaway handle to release the main parachute in the event of an emergency, and a second

Photo courtesy of Brentfinley.com

Figure 5.8   You will learn how to land your parachute.

handle to open the reserve. The cutaway handle is always located on the right side of the harness, where you can easily see it. The reserve handle is always located on the left side of the harness.

A few schools use a special student parachute system with a single operation system (SOS). This rig has just one handle that both releases the main parachute and opens the reserve. The advantage of an SOS system is that it makes it easier for you to deal with a malfunction. The biggest disadvantage is that experienced skydivers never use an SOS system. If you are trained on SOS equipment you will need to learn new emergency procedures to use when you jump your own gear.

All student gear should also have a reserve static line (RSL) that connects the main risers to the reserve ripcord cable, just like a tandem rig. Student rigs also have an automatic activation device (AAD) on the reserve parachute. The RSL and AAD are great backup devices, but you should never rely on them. An RSL and an AAD are required for students by USPA Basic Safety Regulations, but non-USPA centers may skip these important pieces of equipment.

Student parachutes are designed to be forgiving of mistakes. They generally have a forward speed of about 15 to 20 mph, turn slowly, and recover well from overly aggressive control inputs. Student parachutes are designed to open softly. Landings are usually gentle, but the quality of the landing depends to a large degree on how well you fly the parachute. A student rig usually weighs about 30 to 35 pounds and costs about $3000 to $5000.

EXPERT SKYDIVER RIGS

Equipment used by experienced skydivers is similar to student equipment, but is generally much smaller and lighter, and offers very high performance. Parachutes used by experts range in size from about 80 square feet to more than 200 square feet. As with student parachutes, experts select a parachute size based on their weight, with smaller jumpers generally using smaller parachutes. When a heavy jumper uses a small parachute it delivers faster forward speed and a greater descent rate. Some jumpers intentionally select a parachute that is very small for their weight class in order to gain this speed advantage.

Because expert skydivers often use small parachutes, their complete rigs are frequently much smaller and lighter than student gear. There are many manufacturers of parachutes and harness/containers, and prices vary greatly. It is possible to buy a brand-new complete rig for as little as about $2500, or you could pay as much as $6000, depending on style, manufacturer, and options selected. A more expensive rig is not necessarily better or safer.

Second-Hand Gear

If you are thinking ahead to the day when you will be buying your own gear, the complete rig will clearly be your biggest expense. Many students choose to buy a second-hand rig as a way to save money when they graduate from their training. Often recent graduates from skydiving schools purchase a used rig with large main and reserve parachutes that offer limited performance, but are almost as forgiving as the student gear they have been jumping. This used equipment can serve as a bridge for a beginner as the skills needed to fly high-performance parachutes are further developed. These second-hand rigs are often available from other recent graduates, and can be easily sold when you are ready to move on to a new parachute that offers more performance. Many people are justifiably concerned about the hazards of buying used skydiving equipment, but a used rig can serve you well as long as it has been properly inspected by a reputable dealer or parachute rigger.

AUTOMATIC ACTIVATION DEVICES

When you make your first student jump your rig should be equipped with an AAD. The FAA requires this important device on all tandem rigs, and USPA requires an AAD on all other student rigs. It is important that beginning jumpers demand that this life-saving equipment be present on the parachute system that they use.

AADs are electronic or mechanical devices that sense altitude and vertical speed, and are designed to activate your reserve parachute if you are getting too close to the ground and are still in freefall. These devices are designed as backup equipment only, and should never be viewed as an alternative to prompt and appropriate action by a well-trained jumper.

Cypres and Astra

An electronic AAD uses a sophisticated computer to track altitude and freefall speed. It determines when you are in danger by comparing a series of barometric pressure values recorded throughout your skydive. The pressure around a freefalling jumper is constantly changing, and these devices are designed to deploy your reserve only if all the monitored parameters indicate you are too low and still in freefall.

If an AAD fires your reserve at too high of an altitude, it could cause a collision between you and a freefalling skydiver. A mid-air collision between a freefalling jumper and a parachute is a very serious problem. Electronic AADs are designed to prevent these collisions by ignoring rapid changes in pressure that may be caused by body position rather than altitude. Electronic AADs have a computer-controlled self-test function that checks all the circuitry each time the unit is turned on. They also check for required battery power and will not activate if the batteries do not have enough power. Most skydiving centers that use electronic AADs for their student gear

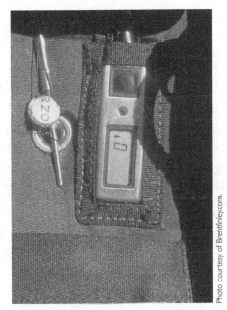

Photo courtesy of Brentfinley.com.

Figure 5.9 A Cypres AAD may be installed on your reserve parachute.

have selected a model called Cypres, manufactured by a German company called Airtech GmbH. (See Figure 5.9.) Cypres is an abbreviation for Cybernetic Parachute Release System. Expert models of the Cypres are designed to deploy a reserve just 750 feet above the ground. The standard Student Cypres should deploy your parachute at either 750 feet or 1000 feet, depending on your descent rate. The Tandem Cypres is programmed to deploy a reserve at 1900 feet.[10] Another model of electronic AAD is called the Astra, manufactured in the United States by FXC Corporation. It is designed to open your reserve parachute at 1000 feet.

FXC 12000

Mechanical AADs have been in use since the late 1960s. A mechanical design uses a pressure sensor coupled with a powerful spring to extract the pin that holds your container closed. These devices are not dependent on batteries or electronic components, and frequently handle rough treatment better than electronic AADs. The disadvantage of a mechanical AAD is that the sensor can sometimes be fooled into deploying your reserve at too high of an altitude if the air pressure around the device is changing extremely rapidly, or if the unit has not been properly maintained. The only brand of mechanical AAD distributed to the civilian market in the United States is made by FXC Corporation, and is called the FXC 12000. This device is designed to open the parachute at an altitude between 1000 and 4000 feet. The specific opening altitude is selected by the instructor on the ground before each jump. A few schools have selected this AAD for their student gear. Some schools like the FXC 12000 because it is less expensive than an electronic AAD, has been on the market for so long, requires no batteries, and is easy to operate.

Maintenance

It is critical that the AAD you are using has been properly maintained. A poorly maintained AAD might not work when needed, or might deploy your reserve when it shouldn't. The Cypres AAD requires new batteries every 2 years or after 500 jumps, and must be inspected by the manufacturer every 4 years. The Astra has no manufacturer-required maintenance or battery-replacement cycle, but a chamber test is recommended each time the reserve parachute is repacked. The FXC 12000 requires a manufacturer inspection every 2 years, and a chamber test each time the reserve is repacked. Since the FXC 12000 is a mechanical device it does not have a computer-controlled self-test function, so the scheduled chamber and manufacturer tests are the only way to be sure it is functioning properly.

You should feel comfortable asking your instructor about the maintenance history of the AADs used by the school, and should be especially curious about maintenance if the school is using the mechanical FXC 12000. Each parachute rig should have a data card that lists the maintenance and test history of the AAD, as well as the history of the reserve parachute. You may ask your instructor to show you this card.

A Cypres AAD will cost between about $1000 and $1200, depending on the model. The Astra is usually a little bit less expensive. The FXC 12000 is about $850.

JUMPSUITS

Jumpsuits can be as simple as coveralls from a hardware supply store, or custom-designed and fitted suits costing many hundreds of dollars.

Most skydiving schools provide their tandem and static line students with a simple jumpsuit or coveralls as protection against dirt and grass stains from landing. Some schools use special skydiving suits for their tandem students because they look better in photographs and video. Many schools will give you the choice to wear a jumpsuit, or let you make your jump without one. It is generally a good idea to wear a jumpsuit as protection against the cooler air at jump altitudes, and to protect you from abrasions on landing.

Freefall students will almost always be provided with a jumpsuit. The specific suit will be selected based on your weight and body type. If you are heavy, a baggy cotton suit will provide lots of extra drag to slow your freefall speed. If you are light, a tight nylon suit will minimize drag and increase your freefall speed. It is important that you and your instructors are able to freefall at about the same speed, and the correct selection of jumpsuits will make that easier. You will also find that a well-sized jumpsuit provides a consistent flying surface that will make it simpler for you to learn basic body stability and advanced freefall maneuvers.

HELMETS

Most skydiving schools provide tandem students with soft helmets called "frap hats," or "French hats." These leather hats look like old-style football or pilot's helmets with extra padding. Tandem helmets are designed to keep you warm in freefall, and protect you from small abrasions or bumps either in the airplane or on the ground. Some students prefer a much heavier helmet, but wearing a hard helmet may be a problem for your tandem instructor, who is right behind you. If you are wearing a hard-style helmet on a tandem jump and move your head back aggressively, you could injure your instructor. For this reason, soft helmets are the only style of head protection that should be used on tandem jumps. Not all schools require or even provide these helmets, and some tandem students and instructors prefer to fly without them.

Students in static line or AFF training are generally provided with a plastic or fiberglass helmet. Many schools use helmets that are commonly marketed for other sports such as skateboarding, kayaking, or rock climbing. Some schools use fiberglass, carbon-fiber, or Kevlar helmets with face shields designed especially for skydiving. A few schools use heavy motorcycle-style helmets. It is important to understand that most helmets used in skydiving have not been tested for safety, do not meet any federal standards, or guarantee any specific level of protection.

Ideally, a student helmet should be lightweight, provide good protection, and should not obstruct or limit vision. If you are a freefall student you will need to be protected against small bumps in the airplane; possible collisions with instructors, risers, and lines moving by your head as the parachute opens; and rough landings. Although motorcycle-style helmets may seem like a good selection, the extra weight can be a burden when the parachute opens, and these helmets sometimes restrict visibility.

Expert skydivers have several dozen models of helmets to choose from, and sometimes even decide to jump without one. Some jumpers like a helmet that keeps all the wind out because it makes the skydive quieter and it is easier to concentrate on flying. Other jumpers like the sound of the wind. Inexpensive helmets cost less than $100, while custom skydiving helmets can cost as much as $400.

GOGGLES

Most skydiving students wear goggles designed for skydiving. Some goggles are very tight-fitting, while others are large enough to accommodate regular eyeglasses. Most goggles are designed with air holes so a small amount of air can move

Figure 5.10 Experts use many different helmet styles.

behind the lens and prevent fogging. There are many different styles of skydiving goggles, and most cost only a few dollars. Skydiving schools will provide you with goggles for your student jumps. If you wear regular glasses or contact lenses, the skydiving school will be able to provide goggles that either fit around your glasses, or goggles that minimize internal air movement to keep your contact lenses in place.

Expert skydivers almost always wear some kind of eye protection. Some jumpers have found they prefer to wear regular tight-fitting sunglasses. They often use a sports strap to hold the sunglasses to their face. Some skydiving schools will let you wear your own sunglasses while you are a student, but you may find that regular sunglasses do not do an adequate job of keeping the wind out. Most instructors will insist that you wear untinted clear goggles or glasses for your freefall training, rather than sunglasses. Dark lenses make it very difficult for your instructor to maintain eye contact with you, and in freefall, eye contact is among the most important ways to communicate.

Some skydivers use custom full-face skydiving helmets with a clear plastic face shield that completely covers their head and makes it unnecessary to wear goggles at all.

ALTIMETERS AND INSTRUMENTS

Skydivers always need to know how far above the ground they are. Altimeters are sensitive instruments that measure air pressure and display it as an altitude reading. Altimeters work very much like the barometers used by weather forecasters, but a skydiving altimeter responds much more quickly to pressure changes, and the display is easier to read in freefall.

Most jumpers wear an analog altimeter on their wrist. (See Figure 5.11.) It has a round face that is about 2 to 4 inches in diameter and is calibrated in either meters or feet. Some jumpers wear these devices on a chest or leg strap. As a jumper descends, the single needle moves counterclockwise to indicate the current altitude. The lower part of the scale is usually color-coded in red and yellow below 3000 or 4000 feet, because that is near standard opening altitudes. Since barometric pressure changes every day, and sometimes through the day, a standard altimeter has a small knob that is used to reset the needle to zero on the ground. Conventional analog altimeters are mechanical devices and do not need batteries.

Some jumpers use a digital altimeter that displays altitude on a liquid-crystal display. These altimeters have a screen about 2 inches wide. Digital altimeters are very accurate but some people have a hard time reading them and interpreting the numbers quickly while in freefall. Digital altimeters require batteries and need to be calibrated on the ground each day. Calibration is usually as simple as turning the unit on and watching the computer display for a moment.

Many experienced skydivers also carry an audible altitude warning device in their helmets. These devices are designed to keep track of freefall air pressure and beep at a preselected altitude. Some of these devices will provide as many as three separate tones at different altitudes. A jumper using one

Figure 5.11 Many skydivers wear an altimeter on their wrist. This altimeter shows that the jumper is at 3500 feet.

of these devices will often set it to beep first at the altitude he plans to start thinking about opening his own parachute. The second tone will frequently be set for the actual opening altitude, and the third tone will be set for an emergency altitude.

Some jumpers also carry altitude-recording computers. These devices are about the same size as audible alerts and usually provide similar tone alerts. They also store detail about freefall speed, freefall time, vertical distance covered, as well as exit and opening altitudes. The small liquid-crystal display on these devices shows basic information about the skydive, but they can also be hooked up to a personal computer to display more detail about each jump. (See Figure 5.12.)

USPA Basic Safety Requirements mandate that all students have a visually accessible altimeter, even for tandem jumps. Most students in tandem, static line, and AFF programs are provided with a conventional analog altimeter. Most experienced jumpers use at least one regular altimeter,

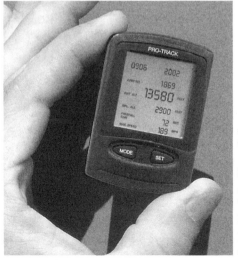

Photo by Adam Buckner/Mercury Grafix

Figure 5.12   The ProTrack is a popular model of altitude-recording device.

and frequently also use an audible device. Most skydiving schools do not provide audible altitude warning devices for student use. It is a good idea for you to develop time and altitude awareness using a simple analog altimeter and your eyes, before you begin to rely on advanced instruments.

Conventional altimeters cost about $150. Audible devices and recording computers can cost as much as $300.

LOG BOOKS

It is always a good idea to keep track of your jumps. All students in static line and AFF programs, and some tandem students, are provided with a log book in which to record important information and comments about each jump. Most log books have spaces in which to record such information as where the jump was made, the type of airplane, type of parachute equipment, exit altitude, freefall time, cumulative freefall time, and how

Figure 5.13 Most jumpers keep a log book.

close you landed to the target. Space in the log book is also provided for detailed comments by your instructor.

As you move through your training program, each new instructor will look at your log book to see what you have done on past jumps, and what skills you should be taught on your next jump. A log book is also a great place to record your experiences and thoughts about each skydive. When you complete your student program the log book will provide proof of your experience and training, and it will be needed to apply for a skydiving license.

Some jumpers stop using their log books after they have a few hundred jumps, while others continue keeping records for thousands of jumps. The USPA issues awards to jumpers who have reached key milestones such as 1000 jumps or 12 hours of freefall time. Accurate log books are required to apply for these prestigious awards, to earn specialized ratings, and to become a skydiving instructor.

Chapter 6

AIRCRAFT

When you make your first jump you may leap from a tiny airplane that holds only five people, including the pilot, or you may make your jump from a larger plane that holds more than 30 skydivers. Key differences among skydiving airplanes include the number of people that can be carried, the number and type of engines, and the size and position of the door the jumpers use for an exit. Although many different types of airplanes can be used for skydiving, a few models have become especially popular among skydivers.

PISTON VERSUS TURBINE

Most small airplanes used for skydiving have conventional piston-powered engines that are very similar to the engines in passenger cars. Some of these piston engines even use regular auto gas. Airplanes with piston-powered engines are relatively inexpensive, often costing as little as $40,000. Piston airplanes, such as the Cessna 182, are ideal for low-altitude flight, but they do not operate very efficiently above about 10,000 feet. Piston airplanes are generally favored by small drop zones

Photo courtesy of Brentfinley.com

Figure 6.1   Skydivers jump from two turbine airplanes over Skydive Arizona.

because the initial purchase costs are low, maintenance is simple, and pilots qualified to fly them are plentiful.

Larger skydiving airplanes often have turbine-powered engines that burn jet fuel. These engines are frequently called turbo, turboprop, or propjet engines. Turbine engines have a standard-looking propeller that is driven by a series of compressors instead of a piston assembly. Turbine engines are far more powerful than most piston engines, and are much more efficient at altitudes above 10,000 feet. These airplanes frequently carry more jumpers than piston airplanes, and they climb more quickly. Many turbine airplanes are retired cargo or commuter airplanes that have been modified for use by skydivers. A skydiving airplane with turbine engines will almost always cost more than $250,000, with some valued at well over $1 million. Maintenance of turbine airplanes is more specialized than piston airplanes, and pilots qualified to fly them require far more training. Skydiving schools that use turbine airplanes such as the

Twin Otter or CASA 212 have invested a great deal of money in their planes, and will almost always promote them in their advertising.

In many skydiving airplanes, the original interiors have been removed to save weight. Often the seats have been taken out and jumpers sit on the bare floor. (See Figure 6.2.) In some airplanes, custom lightweight aluminum benches have been installed. Many jump planes have special doors installed that are easy to open in flight, and some have roll-up doors that occupy very little space and operate easily. Many jump planes also have steps outside the door, and handrails, either inside or outside, so that many jumpers can hold on and leave at the same time. These handrails are called floater bars.

One important FAA requirement is that every person in the plane must be wearing a seatbelt when the airplane is moving on the ground, taking off, or landing. This rule applies to every skydiver, in every airplane. It is often difficult for skydivers

Photo by Dean O'Flaherty

Figure 6.2 In many skydiving airplanes, the seats have been removed.

wearing bulky parachute rigs to wrap a seatbelt all the way around their body, especially when sitting on the floor. Frequently a seatbelt will be mounted against the side wall or floor of the plane and you will need to thread it through your harness, rather than around your body.

The FAA also mandates that airplanes not be over-loaded, and that all the weight must be properly distributed. This is an important safety consideration, and your instructor will make sure you are seated properly, and that there is not too much weight in the airplane.

You will probably find that some skydiving planes look terrible on the outside but are mechanically sound. While the visible condition of an airplane may not be related directly to its mechanical condition, a skydiving center that takes pride in the appearance of its planes will often also be serious about mechanical maintenance. You should be suspicious of a skydiving center that uses old airplanes that appear to be in rough shape.

COMMON JUMP PLANES

Many models of airplanes are used for skydiving operations, and there are many different versions of each model. The primary differences among airplanes of the same basic model are take-off and climb performance. The performance is determined primarily by the type of engines installed on the plane. Airplanes can be fitted with basic engines, or more powerful engines that deliver more power. Most skydiving centers will be happy to tell you about the kinds of airplanes they use, and the performance you should expect in terms of number of jumpers carried and the time to climb to the jump altitude. Some of the most common jump planes are described in this chapter. If you encounter a skydiving center that uses another kind of airplane, the school should be able to compare it to one of those listed here, and give you an idea of how it stacks up against the others.

Cessna 182

One of the most common skydiving airplanes is a Cessna 182. This is a small, single-engine piston airplane that will usually only hold four jumpers and the pilot. Many pilots and jumpers simply refer to these planes by their basic model number, so you may hear a school call this kind of plane a "182." Because the Cessna 182 has a piston engine it is most efficient at lower jump altitudes, so these planes are frequently used by schools with static line programs, and by skydivers who do not jump from altitudes higher than about 10,000 feet. A Cessna 182 will often take about 25 to 30 minutes to climb to 10,000 feet, and much longer to climb higher. The Cessna 182 and its close cousin, the Cessna 180, are relatively inexpensive to purchase and simple to fly. Many have been in service since the late 1950s or early 1960s. They work well on short runways, and even take off easily from dirt and grass fields. Small and rural drop zones tend to use 182s as their primary jump airplanes. Jumpers generally sit on the bare floor of these planes and exit from a right-side door that usually has a hinge on top rather than the side, making it easier to open in flight.

Many schools have students climb out of the 182 and hang on a diagonal strut as part of the exit process. (See Figure 6.3.) This seems very difficult, but it is easy to do, and will actually help you position yourself in the air. You may be afraid that you will not have the strength to hold onto the strut at exit speeds of 80 or 90 mph, but the wind supports much of your body weight, making it simple to hold on. The advantage of a hanging exit is that it positions you in the air stream and makes it much easier to control your body when you leave. Many students who are initially afraid of hanging outside the airplane find it is really fun, and often think of the process as body surfing in the wind.

Figure 6.3   A hanging exit from a Cessna 182 can be a lot of fun.

If you make a tandem jump from a 182, your instructor will connect your harnesses together inside the plane before the door opens. When it is time to exit, many instructors have their student climb out onto a small foot step outside the door. This is an awkward move, but you can expect your instructor to help guide you along the way. Many students find this the scariest part of the jump. If you think you will find this scary, it may help to remember that you have an experienced instructor with you. If you slip off the step your instructor will be right there to help you, and of course you have a parachute and were planning to jump anyway, so slipping off the step isn't a big deal at all.

When a Cessna 182 is used for Accelerated Freefall, the student and two instructors frequently move onto the small step outside the door and leave as a three-person team. Getting yourself into position will be difficult, but you will have a chance to practice on the ground, and your instructors will help you when it is time to do it for real.

Pilatus Porter

A Porter is a small, single-engine turbine-powered airplane manufactured by the Pilatus aircraft company. It is frequently used by medium-sized drop zones. A Porter will carry about 8 to 10 jumpers and climb to 12,000 feet in roughly 30 minutes or less. It has a sliding door on the right side that makes it easy to jump out. A Porter is designed for use on very short runways, and functions well at small airports where many larger airplanes cannot operate.

Cessna Caravan

Larger skydiving centers may use a Cessna Caravan. These airplanes have a powerful single turbine engine, and they work very well at higher jump altitudes. A typical Caravan will carry 12 to 14 jumpers, and climb to 12,000 feet in about 20 minutes. A Grand Caravan is an extended version of the basic model that carries 18 to 20 jumpers and climbs faster. Cessna introduced the Caravan in the mid-1980s, so these planes are all relatively new. Many Caravans are used by small commuter airlines, cargo services, and relief organizations. They do not require long runways, handle well in flight, and carry a big load. Caravans were actually designed with skydiving in mind, so they are very well suited to the sport, and most jumpers really appreciate them.

A Caravan has a large cargo door used for skydiving. Many of these planes have been fitted with a specially designed clear Plexiglas roll-up door, similar to a garage door. Most student exits from a Caravan are easy. The door is large and there is usually no step outside, so students just dive out, making the

Caravan a nearly stress-free skydiving plane. When it is time to jump, the door is pushed up, and away you go.

King Air

The King Air is a twin-engine turbine airplane that generally carries about 14 jumpers and reaches 12,000 feet in about 15 minutes. Some King Airs have been equipped with more powerful engines that cut the climb time in half, making it possible to reach 12,000 feet or higher in less than 10 minutes. The inside of a King Air is usually small. The door in most King Airs is on the left side and is also fairly small, but easy for most people to jump from. Some King Airs have been equipped with floater bars, and many have roll-up plastic doors.

Twin Otter

Twin Otters are also two-engine, turbine-powered airplanes. They typically carry 20 to 23 jumpers and take about 15 to 20 minutes to reach 12,000 feet. Twin Otters are solid airplanes designed for use on short and narrow runways. They are among the best multiengine turbine airplanes for use on grass or dirt runways, climb well, and are relatively easy to fly. You will frequently hear skydivers save a syllable by calling the Twin Otter just an "Otter." Some Twin Otters have been upgraded with extra-powerful engines, and these are often called "Super Otters." Super Otters climb much more quickly than regular Otters, often reaching 12,000 feet in 12 minutes or less. Otters have a large and roomy fuselage that allows small jumpers to stand up inside, but taller skydivers need to crouch down. Most Twin Otters have a large roll-up door and floater bars to hang onto while positioning yourself for the exit. Student exits from a Twin Otter are easy because the door is so tall and wide, and the cabin offers so much room.

Figure 6.4 Twin Otter exits are easy.

Photo courtesy of Brentfinley.com

Skyvan and CASA 212

The Skyvan and the CASA 212 are large turboprop airplanes that offer a tailgate exit. The tailgate is a part of the floor at the back of the plane that can be raised and lowered by the jumpers or pilot. When you board a tailgate airplane the gate is frequently lowered to the ground to form a ramp, and then it is lifted to the closed position for flight. When it is time to jump, the tailgate is opened to a level position, offering a huge platform on which skydivers can stand. The air directly behind a tailgate airplane is usually much smoother than outside a side-door airplane. This makes it easier for jumpers to get stable, and it is especially easy for students.

A Skyvan typically carries about 22 jumpers and climbs to 12,000 feet in about 25 minutes. Some Skyvans have been equipped with extra-powerful engines and can reach 12,000 feet in as little as 9 to 13 minutes. The CASA 212 generally holds about 34 jumpers and also climbs quickly. These airplanes are large enough so that most people can stand up inside. A Skyvan is so big that you can drive a car up the ramp and all the way into the plane. In fact, a few skydivers have actually hitched rides in junk cars pushed out of Skyvans over the Arizona desert. The jumpers ride around for a while, then hop out and open their own parachutes before the cars crash harmlessly into empty desert. (See Figure 6.5.)

Photo courtesy of Brentfinley.com

Figure 6.5 A Skyvan is big enough to hold a car. This outrageous jump was conducted for a spectacular film by cinematographer Joe Jennings called *Good Stuff.*

OTHER AIRPLANES

Skydivers are very inventive, and they are always looking for new experiences and different airplanes to jump from. Most skydiving centers stick to a few basic kinds of airplanes, but sometimes jumpers can find unique planes to leap from.

A few skydiving centers make special arrangements to use helicopters for jumping. These jumps are usually made from only 3000 or 4000 feet, and are more expensive because helicopters cost more to operate. Skydivers like the feeling of a helicopter jump because it can hover, so there is no real wind at exit. When you fall away from a helicopter there is no noise for a few seconds, then the wind begins to build up and gets very loud. When exiting from a helicopter there is no air moving against your body, so some skydivers have a hard time with stability the first few times they jump from one. After about 10 seconds your speed will have picked up to almost 120 mph and it will be just like a regular airplane jump, but those first few seconds are always a treat. Civilian skydiving schools in the United States rarely use helicopters for student training, but some military schools do. For civilians, helicopter jumps are unusual and always fun.

Jumpers can get the same feeling of limited airspeed by jumping from a hot-air balloon. These jumps are especially enjoyable because they often involve a short flight skimming over the local treetops before climbing to higher jump altitude. Balloons will frequently take off from the skydiving center, but they have limited control over where they fly, and just follow the wind wherever it takes them. Skydivers who jump from balloons usually land in small fields away from the drop zone, and then need to hitchhike or find their own ride back.

Some skydiving centers arrange for old biplanes to take jumpers for a ride and a skydive. These airplanes have two sets of wings, one on top of the other, and usually have just two seats and an open cockpit. Sometimes the jumpers will climb out and walk out on the wing, then jump off. Sometimes the pilot will simply turn the biplane upside down and the jumper will push off into a weird inverted freefall. Biplanes are a real treat to fly in, and especially nice to jump from. (See Figure 6.6.)

Skydivers enjoy chasing new experiences, and they are always looking for different airplanes from which to jump. Each type of airplane offers a different exit experience, but after a few seconds the freefall becomes just like any other jump from a conventional skydiving airplane. As a student you will probably jump from one of the common jump planes, but after you complete your training you will have a chance to make jumps from many different types of aircraft.

Photo courtesy of Picture This Photo

Figure 6.6 A skydiver jumps from an inverted biplane above the World Freefall Convention.

Chapter 7

UNDERSTANDING RISK

You may think that skydiving is no more hazardous than an amusement park ride, or you may consider the sport to be almost as dangerous as Russian roulette. In reality, the risk is between those two extremes. If you are like most skydivers, you have an interest in safety but are willing to accept some risk, as long as it seems reasonable and controlled.

Actually, it is difficult to define a specific risk level for skydiving because the sport has so many variables, and there are so few reliable sources of data covering actual activity level. Some skydivers keep accurate records of their jumps, as do some drop zones, but many do not keep any data at all. The annual number of fatalities in the United States is carefully tracked, but it is a low number that varies from year to year, and that tends to skew the data.

Defining the probability of being involved in an accident also requires an understanding of the differences between various kinds of skydiving activities, the experience level of jumpers, and the kinds of problems that may eventually lead to a fatality. Most expert skydivers will not face any of the typical problems associated with a static line failure or tandem malfunction. As a beginning jumper, you won't need to worry about some of the problems unique to the high-performance parachute systems used exclusively by experts, or the specialized jumps they often participate in. (See Figure 7.1.) So, accident statistics really need to be understood in terms of specific problems, as well as the experience level of the jumpers involved.

USPA REPORTS

According to records maintained by the United States Parachute Association (USPA), between 1991 and 2000 there were an average of 33 skydiving fatalities in the United States each year.[11] The vast majority of those accidents involved very experienced skydivers exceeding their own limits. Student

Photo by Dean O'Flaherty

Figure 7.1   Specialized jumps like this "tube dive" are never done by students. This expert skydiver was flying above Skydive Space Center in Florida.

deaths are rare, usually averaging only a few per year, and in many years there are no fatalities involving first-jump students at all. These numbers are interesting, but in order to gauge the actual safety of skydiving it is important to know how many jumps are made each year, and how many jumps are in each category or experience level, and then make some assumptions about specific risk.

The USPA does not receive accurate field reports detailing how many students actually skydive, or how many jumps each person makes. However, the association does have a rough idea of how many experienced skydivers there are, based on its own membership levels. All USPA members are asked to report voluntarily how many jumps they make each year, and some drop zones report details of their activity levels. So, the USPA has some participation data available. The USPA also maintains a database of fatal accidents with probable causes identified. By combining all these data it is possible to build a profile of the relative risks a skydiver faces.

USPA data for 2000 show that there were 32 fatalities, and there were 34,217 members of the organization that year.[17] Using these figures yields a fatality rate of 1 per 1069 members. The fatality numbers vary significantly from year to year, but a 10-year average of data collected between 1991 and 2000 shows one skydiving death recorded for every 903 USPA members. (See Figures 7.2 and 7.3.) USPA single-year data also show that members made an estimated 2,244,165 jumps in the year 2000,[13] presenting a fatality rate of just one death for each 70,130 skydives.

Basic fatality data is interesting, but it doesn't really provide a good guide to understanding why skydivers have accidents, or what you can do to minimize risks. When data provided by the USPA are further analyzed and broken down by categories, some specific problems and risk factors can be identified.

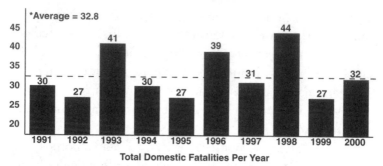

Figure 7.2   The number of fatalities varies from year to year. *(Source: USPA. Graphic by Laura K. Maggio.)*

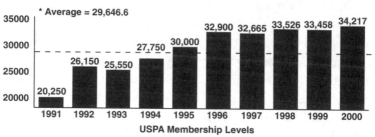

Figure 7.3   USPA membership has grown significantly. *(Source: USPA. Graphic by Laura K. Maggio.)*

The USPA data from 2001 show there were 35 actual fatalities involving skydivers that year, but only 5, or 14 percent, of those who died were classified as students.[14] The vast majority of fatalities, 68 percent, were among the most experienced skydivers,[15] and almost all of the accidents could have been prevented. (See Figure 7.4.) There are several categories of problems that can lead to a fatality, and many of those problems can be minimized or eliminated with appropriate training, equipment selection, or improved procedures. Many of the problems do not apply to students at all.

Understanding how problems occur is the best way for you to learn how to prevent them in the first place. The USPA pub-

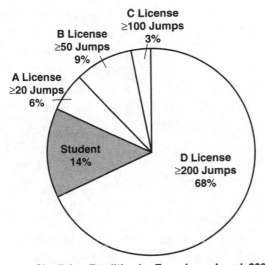

Skydiving Fatalities by Experience Level, 2001

Figure 7.4 Most fatalities claim the lives of experienced jumpers. Student fatalities are rare. *(Source: USPA. License jump numbers were increased in 2003. Graphic by Laura K. Maggio.)*

lishes a detailed report of each fatal accident in *Parachutist*, a monthly magazine distributed free to all members. The organization also publishes an annual summary of the accident data, also included in *Parachutist*. It may seem odd to read about the technical detail of skydiving accidents, but understanding this information is the key to managing risk and preventing accidents.

Specific Problems

Most people tend to think of skydiving deaths being caused when a parachute doesn't open in time, but this is actually a rare problem, accounting for only 5 deaths in 2001, or 14 percent of the total reported by the USPA. In many of these cases, the jumpers either failed to pull the ripcord, or may have had other problems that ultimately prevented them from using the parachutes they were wearing. An automatic activation device (AAD) can often prevent these kinds of accidents by opening the reserve parachute if the jumper is not capable of doing so. USPA Basic Safety Requirements mandate an AAD for all students. Beginners should also be equipped with an altimeter, and should be well trained in maintaining altitude awareness, and trained in emergency procedures. You can easily reduce your risk of a fatality of this type simply by carrying an altimeter and making sure you have been trained to use it.

Parachute malfunctions, or actual failure of a parachute to open properly when a jumper has pulled a ripcord in time, are another cause of death for skydivers, but these too are often preventable. In 2001, seven skydivers died when they experienced unusual malfunctions and were unable to deal with the problems before reaching the ground. Only one of the jumpers in this category was classified as a student. These unusual accidents included an experienced jumper who had a malfunctioning main parachute while making his first skysurfing jump. He became distracted, and then ran out of time before his reserve could open. Another experienced jumper was carrying a long nylon tube that trailed behind him in freefall, and his main parachute entangled with it. He did not follow proper procedures for clearing this kind of problem, nor did he use his reserve in time. Yet another experienced skydiver was wearing a jumpsuit with huge nylon wings that caused instability. He apparently opened his main parachute in an unstable position, allowing it to wrap partly around his body. Likewise, two experienced jumpers had problems when their main parachutes became tangled with their camera helmets. The only student fatality in this category in 2001 involved a man on a static line jump who had a relatively simple parachute malfunction, but he failed to use his reserve parachute in time. You should be trained to open your main parachutes high enough to allow for the use of a reserve parachute if it becomes necessary. Students should always have a higher opening altitude than experienced jumpers, and should never be equipped with extra things like cameras, nylon tubes, skysurf boards, or wingsuits. As a student, your opening altitude will always be above 3000 feet, and it can be as high as 6000 feet when making a tandem or Accelerated Freefall jump.

The largest single category of fatal problems in 2001 was uncontrolled landings, and half of these deaths involved very experienced skydivers jumping with high-performance parachutes. High-performance parachutes are lots of fun to fly because they are capable of reaching speeds of about 80 mph while near the ground. Unfortunately, the high-speed landings that are possible with these parachutes can cause serious injury or death if problems are encountered too low for a jumper to respond. Student parachutes, on the other hand, typically

Figure 7.5   Students should never be equipped with things like the cameras worn by this expert freefall photographer.

have a controllable descent rate of about 11 mph and have far less forward speed than expert parachutes, so landing injuries in the student category are usually far less serious, and are rarely fatal. Landing problems are often caused by a combination of poor training, inappropriate equipment, excessive wind or turbulence, or poor judgment. Almost all the deaths in this category involved skydivers with perfectly fine parachutes simply flying them into the ground. These deaths are almost always preventable through training, equipment selection, and good judgment.

Other skydiving deaths in 2001 were attributable to medical problems, collisions between two experienced jumpers on opening, and even a skydiver who jumped out of one plane and hit a second skydiving airplane that was positioned just a few feet below.

In almost every case of a skydiving fatality listed in the 2001 USPA report, the problems could have been prevented. Requiring beginners to use standard student parachutes with conservative flight characteristics will go a long way toward reducing landing problems, and will make it easier to deal with most kinds of parachute malfunctions. Likewise, the use of a reserve static line (RSL) and AAD can help prevent most problems caused when a student fails to pull a ripcord. All reputable drop zones already adhere to this kind of student training standard.

Of course, even with the best equipment and training, accidents can happen. In 2001 there were a few fatalities that resulted when both the main and reserve parachutes were used at the same time, and became entangled. It is always worth remembering that although the risks inherent in skydiving can be managed, they can never be eliminated.

TANDEM

Perhaps the most interesting statistics are related to tandem skydiving. Tandem jumping has quickly become the most popular form of student training, accounting for about 66 percent of student jumps in the United States, according to numbers compiled by the USPA in 2000,[16] yet tandem fatalities are unusual. (See Figure 7.6.)

When tandem skydiving began in the early 1980s, there were many problems, and fatalities occurred with alarming frequency. The Federal Aviation Administration released data in a 1999 Notice of Proposed Rule Making (NPRM) detailing statistics covering tandem fatalities for the years 1991 through 1996. That data included eight tandem fatalities, with 670,707 jumps made on tandem equipment, or one tandem death for every 83,838 jumps.[17] The FAA data published in the NPRM is old, and does not reflect dramatic improvements to the equipment or training. Many of the weaknesses of tandem systems have been identified and corrected, and now tandem fatalities are very rare. Some of the improvements have involved minor and major equipment design changes. An AAD has been

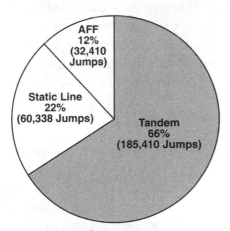

**Total Student Jumps
In 2000 = 278,158
(Based on Instructor Currency Requirements)**

Figure 7.6   Tandem has become the most popular first jump training method. *(Source: USPA. Graphic by Laura K. Maggio.)*

added as a mandatory safety element on every tandem rig, and instructor experience and training have improved significantly. As a result of the changes, statistics reported in May, 2002 by The Uninsured Relative Workshop, a leading tandem equipment manufacturer, show a worldwide 3-year tandem fatality rate of less than one every 420,000 jumps. The data for accidents in the United States are even more impressive, with just one fatality over the same 3-year period, which included an estimated 540,000 domestic tandem jumps.[18]

INJURIES

Although you should of course understand the risks of a fatal accident, you should also think about the possibility of less serious injuries. A bad landing can result in injuries that are sometimes as minor as a pulled muscle, or as serious as broken bones. There is no national database that includes records of student injuries, and most drop zones do not maintain complete records, so it is difficult to determine how many skydivers are injured while participating in the sport. Student injuries are almost always related to poor landings, with the most common injuries affecting the leg and ankle, and occasional compression fractures of the spine. More serious injuries are also possible.

OTHER ACTIVITIES

There is no question that skydivers are exposed to a high level of risk, but the sport tends to draw people who enjoy taking chances, and many skydivers participate in other risky sports as well. Jumpers often try to compare the safety of skydiving with other activities, and to the routine risks all people are exposed to throughout their lives.

The National Safety Council (NSC) publishes an annual summary of fatalities and accident statistics called *Injury Facts*. This short book offers a fascinating statistical look at how people are injured in the United States throughout each year. Although it is not possible to use the NSC data to draw direct comparisons between skydiving and other sports, it is interesting to note how many people die each year from accidents while doing things most of us take for granted.

The NSC data show that unintentional injuries were the fifth leading cause of death in the United States in 1998,[19] and the leading cause of death for those between the ages of 1 and 44. Motor vehicles were responsible for more than 43,000 deaths during the year, falls on or from stairs were listed as the cause of 1389 fatalities, and poisoning was responsible for ending 10,255 lives. Drowning in bathtubs was responsible for 337 deaths.[20]

The annual NSC report highlights data for several sports collected from various sources, but it does not offer enough information about participation levels to compare one activity to another. Boating, for example, was responsible for 734 deaths in 1999, as reported by the U.S. Coast Guard. Snowboarding and Alpine skiing were responsible for 30 deaths during the 1999–2000 season, with 7 snowboarders and 23 skiers killed on the slopes, according to information provided by the National Ski Areas Association. There were 6 fatalities related directly to high school football in the 1999 season, and 3 football fatalities in the 2000 season. The NSC also reported 12 additional deaths indirectly related to football in both 1999 and 2000.[21]

Extreme weather is also responsible for many deaths in the United States each year. In 1999, heat was the cause of 502 deaths, tornadoes killed 94 people, floods accounted for 68 deaths, and lightning killed 46 people.[22]

Data reported for 1998 show that nature can kill in other ways too. Bee stings were responsible for 46 deaths; snakes, lizards, and spiders killed 5 people; and 15 Americans died after being bitten by dogs.[23]

The NSC also maintains data collected from hospital emergency rooms that details injuries not resulting in death. The report makes note that there are significant differences among reporting methods, and the collected data is not always complete, so comparisons among sports should be avoided. Still, it is interesting to read about injury statistics for more common sports. NSC *Injury Facts* reports that 339,775 people were treated in U.S. emergency rooms in 1999 for injuries related to baseball and softball, 372,380 were treated for football injuries, 175,303 were treated following soccer games, 22,639 were treated for bowling injuries, and 2486 were treated for injuries following participation in horseshoe pitching.[24]

RELATIVE RISK

If you are like most people who begin skydiving, you probably already know the sport is potentially dangerous. The perception of danger may even be one of the reasons you are interested in skydiving. Students often begin their training thinking there is very little they can do to control the risk, but quickly come to understand that they have a tremendous ability to help prevent serious accidents. At some point in your training, usually after about 20 jumps, you will begin to lose your natural fear, and begin to feel the risks can be mastered. After a few years, most jumpers gain an awareness of the true dangers and begin to understand that the risks inherent in skydiving can be reduced, but can never be eliminated.

Skydiving schools sometimes point to deaths in other sports or activities as a way of convincing you that jumping out of airplanes is not all that dangerous. It is easy to look at 43,000 total motor vehicle deaths in 2000, and an average of just 33 skydiving fatalities each year, and believe that driving is more dangerous than skydiving. However, you should consider that the actual death rate for motor vehicles reported by the NSC in

Figure 7.7 Skydivers must always be alert for danger.

2000 is 0.156 per 1000 participants,[25] while the fatality rate for active skydivers reported by USPA is about 1.1 per 1000 members, based on 10-year averages.[26] The worldwide 3-year fatality rate for tandem skydiving as reported by The Uninsured Relative Workshop is considerably lower, at 0.0042 per 1000 participants,[27] but that representation can quickly change with just one or two fatalities. So, skydiving can be accurately termed more dangerous than driving, or it can be shown to be safer than driving, depending on how the data are interpreted. You should always be very cautious when reviewing safety data and other statistical detail about skydiving, or any other activity. Statistics can easily be used to tell several conflicting stories at the same time, and should always be suspect.

Fatal skydiving accidents are relatively rare, but they do occur. You should understand that if you are participating in any high-risk sport, an accident can happen, and you

Figure 7.8   Every jump is exciting, and many of the risks can be controlled.

should be prepared for the expense and hardship that may result. Skydiving is an exciting sport, and many of the risks can be controlled with appropriate training, equipment, experience, and well-defined procedures, but the risk can never be eliminated.

Chapter 8

HOW SKYDIVERS FLY

Human flight is pretty amazing, and rarely understood by people that haven't jumped. After you have made your first skydive, you will probably experience the exhilaration that skydivers feel and the addictiveness of the sport. Most people who haven't jumped can't understand that, and skydivers even have a name for them. Nonjumpers are called "whuffos." It's a name originally developed for people who lean against the airport fence with a dumb expression and ask, "What fo' you jump out of them airplanes?" Whuffo. Sometimes they will even shake their heads, and with a condescending snicker repeat the tired old adage, "There are only two things that fall out of the sky...bird crap and idiots." Skydivers hate that expression because it fails to indicate that there is much more to skydiving than falling out of the sky. Most whuffos just don't understand what skydivers can actually do in flight. Perhaps they are too afraid to give skydiving a try themselves, or they may just be so set in their ground-loving ways that they can't imagine what is happening in the sky over their heads. If they could, they would quickly understand that skydivers don't "fall out of the sky," skydivers actually fly, like birds and airplanes.

Once you understand how much control skydivers have, you will begin to see that we really do fly.

ARCH

When you are trained for your first jump you will learn how to present your body to the wind for maximum flight control. The basic position, called an arch, begins by facing into the wind with your arms out at about a 90-degree angle, and elbows bent so that your hands are roughly even with your ears, and a little farther apart than shoulder width. Your legs will be bent at the knees with your feet up higher than your body, and toes slightly pointed. Your shoulders will be lifted back, and your pelvis pushed down. The idea is to establish the trunk of your body as the center of gravity, and use your arms and legs for balance. If you are in a perfect arch position you will be flying straight down, belly first. Think of a badminton shuttlecock that always lands with the rounded heavy part down. When you are arched, your pelvis will be the round heavy part, and your arms and legs will serve as symmetrical outriggers to keep you nicely balanced. (See Figure 8.1.)

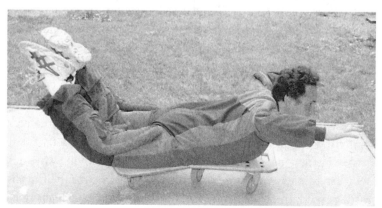

Figure 8.1　An arch is the foundation of flight.

The arch is a basic flying position that is used as a foundation for stability in student training and advanced skydiving. Many schools teach static line students a "hard arch." This places your body in a more pronounced position, with your arms and legs straight out forming a giant "X," but your pelvis is still pushed forward. (See Figure 8.2.) The hard arch is generally used only for the first few jumps of the static line program, and as you build experience it is "softened" to the more common "relaxed arch" position.

If your first jump is a tandem skydive, your instructor may have you leave the airplane holding onto your harness. Your pelvis will still be pushed forward, and your legs will be

Photo courtesy of Archway Skydiving Center

Figure 8.2   A hard arch is often used for static line exits.

bent back. That position will keep your arms out of the way for the exit as your instructor uses his arms to define the upper part of the arch. Soon after leaving the airplane your instructor will have you move your arms into a better arch position, and then you will both be flying together.

Accelerated Freefall students spend a great deal of time learning and practicing a relaxed arch with a modified arm position, sometimes also called a "lazy w."

Many different variations of the arch are used for training. However, the principle of defining your pelvis as the lowest point and using your arms and legs as outriggers remains the same in all training programs.

If your arms and legs are in a perfectly balanced position you will be falling straight down toward the ground. Anything that changes balance or symmetry will cause you to turn or move. For example, reaching out with one arm, or extending one leg, will cause a turn. Extending both legs will cause your body to pitch head down and fly forward. Dropping one knee will cause a turn, while dropping both knees will cause you to move backward. Turns can also be created by dropping a shoulder or hip, or by twisting your body or pushing a leg out. Reversing the standard arch by lifting your pelvis up, instead of pushing it forward, will often cause you to flip or roll, and then fly with your back toward the ground.

As you advance in training you will learn to modify the arch and use your arms, legs, and torso to make all kinds of very specific turns, and to vary your rate of descent. By adding small amounts of forward and backward motion you will be able to make very precise turns that allow you to pivot around your shoulder, knee, hip, chest, belly, or any other specific point. It is also possible to do back loops, front loops, and barrel rolls. You will really enjoy flying your body in freefall and using body movements to control yourself in the high-speed air.

FLYING THE PARACHUTE

When the parachute opens, things slow down a bit, yet you remain in complete control of your movement. The parachute is really an inflatable glider with as much control as a hang glider or powerless airplane. Some jumpers like to make their parachutes go really fast, while others like to linger in the sky. Some skydivers like hard and steep turns, while others like smooth turns at slower speeds. Parachutes designed for beginners tend to be very slow and forgiving, but you will find they offer plenty of performance, and the ability to maneuver wherever you want to go.

When you have completed your student training, after about 20 jumps, you will be ready to start experimenting and trying even more exciting things in the sky. Most skydivers develop skill in a skydiving specialty that suits their personality. There are many dramatically different ways to fly, and they all appeal to different people. Each specialty takes advantage of the aerodynamics of flight to provide jumpers with unique ways to fly their bodies and parachutes.

FORMATION SKYDIVING

Formation skydiving is one of the oldest forms of group skydiving. It is sometimes called relative work, or RW, because jumpers move their bodies in freefall relative to each other. Formation skydivers fly on their bellies, just like students, but the similarity ends there. Precision and control is the name of the game for formation skydivers, who measure their movement in inches and fractions of a second. Skilled jumpers can build small formations of just a few people, or giant formations with hundreds of skydivers all joining together.

When two jumpers get together in freefall it is called a "two-way." When four jumpers get together it is called a "four-way." Expert skydivers often compete in four-way and eight-way events. Skydivers in these competitions build a series of very specific formations with names such as donut, cat, zipper, and accordion. (See Figure 8.3.) There are dozens of recognized formations that experienced formation skydivers know how to build.

Often jumpers build larger formations, sometimes including hundreds of skydivers at one time. Building these giant formations requires several big airplanes that start off by flying in formation themselves. Jumpers leave the airplanes, which are flying wingtip to wingtip, and converge together in freefall to build a very specific freefall formation. These huge jumps sometimes start from about 18,000 feet, giving the skydivers just over 90 seconds to get out of the airplanes, build the formation, then move far enough apart to open their parachutes without colliding. Building these huge formations is extremely challenging and requires amazing flying skill and organization.

Photo courtesy of Brentfinley.com

Figure 8.3   Formation skydivers build a zipper.

Formation skydivers generally wear tight jumpsuits with "grippers" on the arms and legs. The grippers are like handles that jumpers use to hold onto each other so the formation has the intended shape. Skydivers in competition are required to define the formation by taking grips at very specific places on the arms and legs, so the grippers help them to build formations according to the rules.

Formation skydivers are often accompanied in freefall by photographers who fly above or below the formation while wearing cameras on their heads. In competition, the video is judged to be sure the formation is built correctly, and that all the grips between people are complete. Even noncompetitive formation skydivers rely on video to provide a broad view of the formations they build.

Creating formations in the air requires lots of practice and carefully choreographed movement. When you are at the skydiving center you will probably see formation skydivers practicing their moves on the ground in their jumpsuits. Often they will start the practice standing up, then actually lie down and create the formations on "creepers," which are small wood or fiberglass platforms with wheels. These creepers allow them to closely replicate the moves they will make in the air. (See Figure 8.4.)

Formation skydivers learn the basics of belly flying as beginning students on their very first jump. In fact, if you are an AFF student you will begin by flying your body in a three-person formation, with your instructors holding onto you. After a few jumps your instructors will let go and you will be flying relative to them without grips. The training you receive at the end of an AFF or static line program will include techniques for leaving an airplane several seconds after your friends, and then catching up to them in freefall. These skills are the foundation for advanced formation flying.

Figure 8.4 Skydivers practice building a "donut" formation on creepers.

FREEFLY

Not all skydivers fly on their bellies in a classic arch. Several years ago skydivers began flying in a seated position, then transitioned to a head-down position that actually involves flying upside down. These flyers regularly move from a belly position to a seated position, to a head-down position, to a standing position, and frequently transition by doing cartwheels or other wild moves in the air. Every time they change their orientation in the sky they change the way their bodies are presented to the air, and change the way they maintain stability. Arms and legs still provide stability and propulsion, just as in the classic arch, but the way the arms and legs are used changes with each position. (See Figure 8.5.)

These skydivers also fly in groups relative to each other, but their movements are in both a horizontal and a vertical plane. Skydivers flying on their heads generally move at speeds

Figure 8.5 Freeflyers enjoy flying upside down over Skydive Space Center in Florida.

of about 165 mph or faster, and seated jumpers often move at speeds closer to 140 mph. By rapidly changing speeds these jumpers can move over and under each other, and they can slide sideways, creating a very wild three-dimensional dance. Slight changes in body position create large changes in speed and horizontal movement, allowing freeflyers to build formations with jumpers in all kinds of odd positions.

The dramatically different speeds that freeflyers are capable of achieving on the same skydive make the threat of freefall collision a critical concern. If a flyer moving at 140 mph slides under a jumper moving at 165 mph, the collision will have the energy of the 25-mph difference in speed, and is very serious. A collision between a belly flyer and a head-down jumper would be even more serious. Freeflyers deal with this problem by jumping in smaller groups than classic formation flyers, and by maintaining constant vigilance all around the sky.

Freeflyers often use baggy jumpsuits with lots of extra material to give them added drag and control at the high flying speeds. Many freeflyers wear helmets with small video cameras attached, so they can review the tapes on the ground and see what their formations and movements looked like. Freeflyers practice their maneuvers on the ground, just like formation flyers, but generally only while standing up. They must be able to visualize how their movements will look and flow together in the air without actually building the intended formations in a seated or head-down position. The prejump planning stage is especially important to help freeflyers create a dance that avoids freefall collisions.

If you are interested in freeflying you should complete a standard student training program and have a good understanding of belly flying. Then take a few lessons in freeflying and learn to freefall under control in a seated position. From that position you can add head-down and other positions, as well as the advanced transitions used by freeflyers.

SPEED SKYDIVING

Some skydivers really like extreme speed. A jumper in a head-down position moves much more quickly than a jumper in a classic arch position because less surface area is presented to the wind. Speed skydivers take advantage of this by flying in a head-down position, and then pulling their arms and legs very close to their bodies. By pulling their arms and legs in tight they are able to reduce drag and increase speed, but that also has the effect of decreasing freefall stability. If even a hand is out of place, the speed skydiver will start to oscillate wildly, and freefall speed will be dramatically reduced. Skydivers who are making speed jumps need to be exceptionally strong flyers and comfortable with controlling stability before they begin serious attempts at high-speed flight.

Speed skydivers generally wear skintight suits without any grippers, often making use of slick suits made for ski racers, or custom suits designed for other high-speed body sports. The rigs worn by speed skydivers are often aerodynamically designed and sometimes integrated into custom suits that smooth out any bumps. Speed skydivers will try anything to gain a few extra miles per hour of speed.

Sensitive altitude computers that also serve as audible altitude warning devices measure skydiving speed. Some jumpers wear these devices in aerodynamically designed helmets, while others wear them on their ankles or wrists. The air

Photo by M. Lillard/Mercury Grafix

Figure 8.6   Some skydivers enjoy flying as fast as possible.

pressure around a jumper's head tends to change rapidly and in unpredictable ways, so jumpers often use speeds recorded by freefall computers worn on their legs or arms. The record of a jump in a freefall computer can be downloaded to a regular personal computer. Once the information has been captured in a personal computer the skydive can be plotted on a graph showing altitudes and freefall speeds broken down by fractions of a second. The plots will show clearly when the peek speeds were recorded, and also offer average speeds over several time periods.

The best speed skydivers are able to fly their bodies straight at the ground at more than 300 mph. These exceptional speeds are extremely dangerous. If a parachute accidentally opens at such high speed it could cause a very serious injury, so speed skydivers are very focused on maintaining their equipment in top condition. The high speed also means a jumper is descending unusually quickly, so it is especially important to keep track of altitude so that the skydiver can transition to an arch position and slow down in time to open the parachute at a normal altitude and speed.

WINGSUIT FLIGHT

Not all skydivers want to fly super-fast. Some like the feeling of cruising across the ground like an airplane, covering several miles on a single jump. Large wingsuits with extra material between the arms and body, and between the legs, allow jumpers to slow their decent rate to less than 40 mph, and increase their horizontal movement. The material is designed in two layers with air pockets, so it inflates and creates a wing surface between the jumper's limbs. By stretching out their arms and legs to catch the most air possible, these jumpers are able to dramatically extend their flight time and range. They can also make sweeping turns like a banking airplane.

Skydivers wearing these advanced wingsuits can fly in groups, and sometimes look like a flock of birds way up in the sky. (See Figure 8.7.)

Some wingsuit jumpers even fly alongside the airplane they jumped out of. This requires a skilled airplane pilot, because the plane needs to match the high descent rate of the wingsuit flyer, and the pilot needs to be careful not to collide with the nearby jumper.

Wingsuits come in a variety of styles offering different performance features. A typical wingsuit will be custom-built and cost $300 to $650. Jumpers wearing wingsuits often measure their descent rate with altitude-recording computers, and sometimes record the distance they cover using Global Positioning Systems (GPS) that track movement by using satellite signals.

Photo by Craig O'Brien

Figure 8.7 Large wingsuits allow a sky-diver to reduce the descent rate and travel great distances across the ground.

SKY SURFING

Sky surfing is an artful form of skydiving that involves freefall while flying on a small board. The boards vary in length from just a couple of feet to about 5 feet long. Sky surfing boards are usually made of wood, fiberglass, graphite, or other light-weight material, and have special bindings that hold the jumper's feet in position. The boards are generally set up with both feet facing forward, one in front of the other. (See Figure 8.8.) The bindings have a release line that is attached to a handle that jumpers usually place near their waist. If there is a problem in freefall, or with the parachute opening, the board can be quickly released. Some sky surfing boards have a small round parachute attached that will open automatically if the board is released while in freefall. Jumpers flying in crowded or urban areas sometimes use these emergency para-chutes on their sky surfing boards. The sky surfing board is also designed for quick release just before landing. Generally, a sky surfer will deactivate the emergency system on the board

Figure 8.8   A sky surfer cruises over the ground on a sky surfing board.

and then release it inches above the ground, just seconds before touching down.

The sky surfing board presents a jumper with a huge surface area to maneuver in the air. Sky surfers use their arms and bodies to position the board aerodynamically, and use the wind resistance and lift generated by the board to move around the sky. Sky surfers can do wild tricks such as cartwheels and spins.

Sky surfing is especially dangerous because the board has such a dramatic effect on flight and can tangle with the parachute at opening. For his reason sky surfers need to be expert skydivers before they ever try flying a board.

Special training in the use of skysurfing boards is available at some large drop zones. Many of these drop zones have formal training programs and will even rent the special equipment needed. Generally, you should have several hundred regular jumps and be a good freefly jumper before trying skysurfing.

HIGH-ALTITUDE JUMPING

Most skydiving takes place at altitudes of no more than about 14,000 feet above the ground. As an airplane climbs higher than that, the air is thinner and has less oxygen available. When skydivers are trying to make really big formations, or extend freefall, they will sometimes jump from altitudes of about 18,000 feet, and experience freefall times of roughly a minute and a half. These jumps always require extra coordination and support, and the skydiving center will provide oxygen in the airplane, usually delivered to each jumper through a mask.

Sometimes skydivers want to jump from even higher altitudes, and a few skydiving centers will accommodate them. Several drop zones schedule jumps from 21,000 feet, and one California drop zone takes jumpers to 30,000 feet, but only on rare occasions. Jumps from 30,000 feet require special training that includes experience in an altitude chamber to simulate the low-pressure environment that will be encountered at high altitudes. Although this advanced training can be helpful when jumping from lower altitudes, it is essential when jumping from 30,000 feet. The air is so thin at 30,000 feet that a jumper without oxygen could easily pass out in less than a minute. There is also a danger of the nitrogen that is always in your blood forming bubbles and causing a serious problem called decompression sickness, or "the bends," a threat that is sometimes encountered in scuba diving. High-altitude skydivers deal with these threats by prebreathing oxygen for about 30 minutes before the plane even takes off, then continue breathing oxygen all the way up. They also wear oxygen masks while in freefall. The prebreathing of 100 percent oxygen reduces the nitrogen in the blood and helps prevent the bends.

Jumpers at high altitudes will also encounter much colder air. The temperature at 30,000 feet often hovers around 40 degrees below zero, so skydivers must wear extra clothing to prevent frostbite. Typically a high-altitude jumper wears several thermal layers, a heavy jumpsuit, boots or heavy shoes, a full hard helmet modified to hold an oxygen mask, goggles, and sometimes an extra face covering for protection against the cold.

A typical freefall from 30,000 feet lasts about 2½ minutes and costs several hundred dollars. The preparation and cost deter many skydivers from ever trying high-altitude jumps, but skydivers who do make jumps from those altitudes really enjoy freefall in airspace that is usually used only by commercial jets.

PARACHUTE SWOOPING

Skydivers do not limit their fun to freefall. A modern high-performance parachute, also called a canopy, is capable of moving at about 80 mph, and that speed can be quite a rush when near the ground.

Experienced skydivers often enjoy diving their parachutes toward the ground to build up speed, or making fast turns to add speed. When they get near the ground the steering toggles or risers are pulled down to level the parachute out, and allow the jumper to fly inches above the ground while bleeding off the speed. These radical approaches are called "swoops," and are very impressive to watch. When you are at the skydiving center you will almost certainly see jumpers swooping inches above the ground for great distances. When done correctly, swooping is an amazing sight that really defines the flight performance of modern parachutes. Some skydiving centers have harnessed the excitement of swooping by organ-

Photo by Donna Marshall

Figure 8.9 Skydivers enjoy high-speed flight over water-filled ponds.

izing competitions that are based on speed, distance covered, or style. Some of these competitions use windblades or flags as gates, and the jumpers fly their parachutes around these obstacles to demonstrate control. Other drop zones have built ponds filled with water so that jumpers can swoop their parachutes while kicking up a spray as they go by. (See Figure 8.9.)

Swooping is breathtaking to watch, but it is also extremely dangerous. The top canopy swoopers use tiny parachutes, sometimes measuring less than 80 square feet, and fly them at tremendous speed very close to the ground. A miscalculation can quickly end in disaster. Although many experienced skydivers may try a bit of swooping just for fun, organized swoop competitions are frequently limited to the very best parachute pilots.

CRW

Some skydivers like flying their parachutes in tight formations and actually grabbing onto each others' canopies. Since the parachutes are flown close together, or relative to each other, the activity is called "canopy relative work," often abbreviated as CRW and pronounced "crew."

CRW jumpers leave the airplane and open their parachutes immediately, then fly together and build formations with other parachutes. The parachutes used for CRW can be regular parachutes like most jumpers are using, or they can be modified and designed just for CRW. Parachutes that are designed to be flown together generally fly a bit more slowly, and are more stable than regular parachutes. The lines that connect a CRW parachute to the jumper are often designed to be easier to grab and hold onto.

Jumpers build a vertical canopy formation called a "stack" when one parachute pilot actually bumps a parachute against the back of another jumper, who then grabs the nylon material and holds on. The higher jumper can then work his or her way down the lines and slip his feet into risers just above the lower jumper's head. This stacking of parachutes can continue as more and more jumpers join the group. (See Figure 8.10.) Jumpers can also build diamond-shaped parachute formations or other shapes by flying their parachutes precisely together and taking grips in different ways. The world's best CRW flyers have built parachute formations with more than 50 jumpers holding on at one time.

If the parachutes in a formation are not flown correctly, or the docks between jumpers are too hard, one, both, or all of the parachutes in a formation can collapse and tangle. These entanglements are called a "wrap" and are very serious. CRW

Photo courtesy of Brentfinley.com

Figure 8.10 Jumpers can maneuver their parachutes together and fly them as a stack. This stack was built over San Carlos, Mexico, by a team called Plaid Jacket.

jumpers always need to be ready to release their parachutes if a wrap occurs, and usually fly together only when they are high enough to allow use of their reserve parachutes if something goes wrong.

EXHIBITION JUMPING

Some skydivers enjoy jumping at big public festivals or events that happen away from drop zones. These jumpers are called exhibition jumpers, or demonstration jumpers, often shortened to demo jumpers. They usually have lots of experience and have trained for many years to make these exceptional skydives. Demo jumpers generally have made hundreds or thousands of skydives before they start making jumps away from the drop zone. Most begin by jumping into local festivals in big fields near their home drop zones. When they can handle those simple events easily, they begin making jumps into smaller areas, or into stadiums filled with spectators.

Exhibition jumps are especially difficult because they are often made into urban areas or stadiums with strange or turbulent winds, or into very small landing areas crowded with spectators. Exhibition jumpers also frequently carry giant flags that trail behind their parachutes, or smoke canisters to help the crowd see them. These extra pieces of equipment can cause problems that need to be considered when the jump is planned.

BASE JUMPING

Not all jumpers need airplanes. BASE jumpers make their jumps from fixed objects, a very risky form of jumping. BASE is an acronym that stands for Building, Antenna, Span (bridge), and Earth (cliff). These are the four object groups from which BASE jumpers like to leap. (See Figure 8.11.) BASE jumping got its start in the late 1970s and early 1980s when skydivers began jumping from El Capitan, a giant 3000-foot cliff in

Figure 8.11   BASE jumpers enjoy leaping from fixed objects like this cliff in Ontario, Canada.

Yosemite National Park. After successful jumps from El Capitan they began jumping from other objects, and started an organization called the United States BASE Association (USBA) to help develop the new sport. BASE jumpers often like to say that their sport is not really skydiving because they don't jump from an airplane, and the USPA has nothing to do with BASE jumping.

Jumpers who make at least one jump from each type of object can apply for certification as a BASE jumper, and receive a numbered award. More than 700 jumpers have completed the four-jump sequence and earned a coveted BASE number.[28]

Until recently there were not many places where BASE jumping was allowed, so jumpers often climbed objects clandestinely, and then tried to flee without being caught. If they were caught jumping from things like downtown skyscrapers or TV antennas, BASE jumpers would often face charges of criminal trespass or reckless endangerment. Because so many of their jumps were illegal, BASE jumping developed as a secretive outlaw sport. BASE jumpers would sometimes identify themselves to each other only by secret number, and were very careful about who they told of their exploits.

In the early 1980s jumpers received permission to hold an annual BASE festival at the New River Gorge Bridge in West Virginia. The bridge is about 876 feet tall where it crosses over a river, and is an ideal BASE site. Between 300 and 400 jumpers travel to West Virginia for the event each October. The bridge is normally closed to pedestrians, but on that day as many as 100,000 people usually walk out onto the bridge and watch the BASE jumpers.

BASE jumpers have also been able to make legal jumps from cliffs in the western United States, as well as a few other domestic sites. There are also a few buildings, bridges, antennas, and cliffs in other countries that attract BASE jumpers.

Most BASE jumpers have hundreds of skydives from airplanes before they jump from any object, but a few people have made BASE jumps without any conventional skydives. If you are interested in this very risky aspect of the sport, there are a few BASE jumping schools that will provide the needed training and equipment. You should understand that BASE jumping is extremely dangerous, and even the most cautious jumpers can be hurt without warning.

MIXING IT UP

When you make your first jump you will be focused on learning basic stability so you can fall straight down while under

Photo courtesy of Brentfinley.com

Figure 8.12 Learning advanced skills will allow you to do wild tricks like Greg Gasson and Omar Alhegelan, who are flying over Skydive Arizona.

control. As you gain experience you will learn to turn and do other simple tricks. As you learn new skills you will feel like you are flying your body and not just falling. Most skydivers don't stick with one discipline, but try many. You will find that skydiving is so much fun because there are so many different ways you can fly your body and your parachute.

Chapter 9

ALL ABOUT DROP ZONES

E very drop zone is different, often reflecting the personalities of the owners and the skydivers who spend time there. Some drop zones are huge operations focused on generating income; others are just tiny collections of friends who have been jumping together for many years, and organized the drop zone so they would have a place of their own to skydive. Often the drop zone, or DZ as it is frequently called, becomes the center of a skydiver's recreational life. Skydivers find themselves hanging out at the DZ even when they are not jumping, and close friendships develop among these jumpers and their families.

ORGANIZATION

Some drop zones are organized as clubs, with each member taking an active role in supporting the operation, while others are large companies with a paid staff of professionals. In some cases the skydiving school is a core part of the DZ, while in other cases it may be a completely separate business just renting space on the drop zone. You may even encounter a drop zone with several competing schools that share airplanes and a landing area, but are otherwise independent. The specific business relationship may be transparent to you, or the skydiving school may clearly promote itself as a separate club or business.

Opening a DZ is pretty easy, but keeping it running requires a lot of work. The most difficult part of starting a DZ is often finding an available airport. Ideally, a drop zone will be located close enough to a city to draw student business, but far enough away that busy air traffic is not a problem. The drop zone needs to have adequate space for airplanes to take off and land, and for parachutes to land. The DZ property can be as simple as a farmer's field with a small grass runway and a trailer to serve as an office and classroom, or it can be a busy airport with multiple long paved runways and many private buildings. The type of airport will often dictate the growth potential of the skydiving business, and help define the character of the operation.

Many drop zones in the northern United States are seasonal businesses, shutting their doors over the winter. A few northern DZs continue operating through the cold winter, but at a slower pace. Most southern DZs are more active in the wintertime, and then slow down when the summer temperatures become uncomfortably hot. You may find that some of the people working at the drop zones in your area are seasonal workers, traveling south for the winter and spending the summer at a northern drop zone.

Large metropolitan areas often have many competing drop zones. Some of these DZs are very small, while others are huge businesses with dozens of instructors and staff. Large drop zones tend to be located near cities such as Boston, New York, Philadelphia, Orlando, Dallas, Los Angeles, Phoenix, and Chicago. Smaller drop zones can be found in all parts of the country, including large urban areas, where they may be located close to a larger skydiving center. In many rural parts of the

Photo by Mike Lanfor

Figure 9.1   Some skydivers jump all winter long in spite of the cold.

country, the only drop zones available are small single-airplane operations run by groups of friends.

SPECTATORS

Most skydiving centers have a special area set aside for spectators to watch the jumping activity. These areas are usually designed to keep observers out of the landing area, but still give them a chance to see all the action. A few spectator areas have picnic tables and room for children to wander freely. Some drop zones invite spectators to walk around the DZ and talk with jumpers, or watch them practice for their jumps. Students and accompanying family members are usually given extra leeway to enjoy everything the drop zone has to offer.

Many students bring a still or video camera along to record everything that happens on the ground. You will not be able to take the camera on your actual jump, but most schools can arrange to have a professional skydiving photographer accompany you in freefall and document the entire experience. Freefall photography usually costs less than $100, and includes some coverage of your training as well as the actual jump. Some students request video coverage on a freefall jump as a training aid, while many tandem students ask for video and still photos as souvenirs.

SOCIAL ACTIVITY

When you visit a drop zone, you will discover that skydivers like to jump a lot during the day, then socialize at night. There are frequently small parties or other activities in the evening, with some rural drop zones hosting nightly bonfires or other outdoor events. Often, skydivers share rides to the drop zone, and sometimes even share vacations to other skydiving centers. Skydivers almost always build great friendships away from the drop zone, and find themselves visiting each others' homes and sharing family events.

Skydiving students frequently become friends with other students who are beginning their training at the same time. These friendships are great for providing emotional support and sharing the thrills and fears of the early jumps, and often last for many years.

Drop zones on private airports or in rural areas sometimes have camping areas available so visitors can spend the night inexpensively after a day of jumping and an evening of socializing. Drop zones without camping areas usually have a list of nearby hotels, and may have arranged discounts for jumpers and students.

Just about every drop zone holds huge parties several times a year. These parties, called "boogies," usually feature special skydiving events coupled with nighttime festivities. Boogies are a chance for skydivers to jump hard throughout the day, and then unwind with their skydiving friends and families in a social setting at night. Boogies are frequently well advertised in skydiving newsletters, and draw jumpers from other drop zones for extra camaraderie. The added influx of

Figure 9.2 Some drop zones have camping available.

jumpers also gives skydivers a chance to make bigger formations than usual, and then swap new stories in the evening. Boogies are generally concentrated around holiday weekends such as Memorial Day, Independence Day, Thanksgiving, and Christmas, and sometimes feature competitions among the jumpers in various skydiving specialties.

Beer Rules

Many drop zones take a special interest in helping their students assimilate into the DZ culture. Most DZs have developed an informal policy of "beer rules" that say a skydiver who has done something exciting for the first time should buy some beer and share it with the other jumpers. The informal rule usually doesn't apply to your first jump but can be invoked for just about any other milestone, such as graduating from the student program, your first four-way formation, or your first time jumping a new parachute. Beer rules also apply when a jumper does something scary or foolish, such as opening a parachute dangerously low, or landing in a hazardous part of the drop zone. The idea behind "beer rules" is to encourage students and beginners to hang out after skydiving and share their experiences over drinks with more seasoned jumpers. Some jumpers substitute home-made cakes or cookies, or bring along something else to share while socializing. Beer rules are a tradition among jumpers that serve to bring everybody together at the end of the day.

Drinking in the evening has long been a part of skydivers' social scene, but skydivers do not drink alcohol in the daytime. In fact, most drop zones have very strict rules that prohibit any open containers of alcohol at the airport until after the last jumpers have landed for the day. This rule almost always applies to spectators as well as jumpers. If you are planning to make a skydive, you should not drink any alcohol for at least 8 hours prior to a jump, or use any drug that may have an intoxicating effect. If you or your friends bring alcohol to a drop zone for a picnic, you should check with the drop zone staff for local rules, and respect their policies.

LARGE DROP ZONES

Many students begin their training at big drop zones. These skydiving centers often fly multiple-turbine airplanes such as the CASA 212, Twin Otter, or King Air, and have a full-time staff of instructors with lots of experience. Large drop zones are frequently 7-day-a-week operations, with jumping starting early each morning and continuing until sunset. They are almost always located near a major city, yet far enough outside the urban area to avoid congested traffic patterns and commercial air traffic maneuvering to land at city airports.

Large skydiving centers usually offer students many training options, frequently focusing on tandem and AFF. Some of these drop zones have advanced training programs such as Skydive University available, and most have regular customers who like to do formation skydiving, freestyle, sky surfing, canopy swooping, and other skydiving specialties. Some large drop zones have separate schools with formalized training programs to help you learn these advanced techniques.

Large drop zones often have retail stores that sell skydiving equipment and souvenirs such as hats and T-shirts. Many skydiving stores also have parachute rigging services available, so jumpers can easily have their equipment repaired and maintained. Large drop zones frequently have a snack bar or informal restaurant, and some even offer day-care services for the children of regular jumpers.

Figure 9.3 Many large drop zones have equipment dealerships on site, such as The PROshop, at The Ranch Parachute Center in New York.

When large drop zones host a boogie, it is usually based around a special skydiving event such as a huge seminar, or the addition of extra airplanes so jumpers can build giant formations. The nighttime parties that follow jumping often feature live bands, and go on long into the night.

Checking In

If you visit a large drop zone you will probably need to check in at the school or a manifest station when you arrive. Manifest is a place where experienced skydivers sign up for their jumps. When skydivers are ready for a jump, they approach manifest and place their names on a list. When there are enough people to fill the airplane, the group takes off. At large drop zones the manifest station is often the hub of activity and is frequently staffed by a full-time employee. When you first arrive at a large drop zone, look for a sign that will direct you to the school office, or seek out a confident-looking jumper for directions. If all else fails and you can't find the school, look for the manifest station.

Large skydiving centers sometimes handle several hundred tandem students and dozens of advanced students over a typical weekend. Some of the most active training programs help more than a hundred new jumpers earn their skydiving license each year, and many of these students go on to advanced training in specialized skydiving disciplines. One advantage of such a large school is that the staff will be extremely experienced, and should have a very well-defined training program.

A large drop zone offers many training advantages, but it sometimes takes a special effort on your part to fit in and make friends. A few students sometimes feel lost in the hustle and bustle of a busy drop zone, and sometimes the staff can become overwhelmed, and may seem inattentive. Of course, since a large drop zone has so many people, and so many students, you will probably find other people to share the experience with.

Most large skydiving schools are open 7 days a week. This makes it easy to fit skydiving training into your schedule. Many students find big drop zones that are very busy on weekends become much quieter places on weekdays. Often instructors will be able to spend more time with you during the week, and the overall feeling of the drop zone will be more comfortable on a weekday. Some students also take advantage of weekday operations to make a few jumps after work. Weekday students may have jobs with odd hours or rotating shifts like police work or firefighting, and they especially appreciate the flexibility offered by a full-time drop zone.

Running a large drop zone is a monumental task. These DZs often employ dozens of workers, many of them full-time employees. Most large drop zones also rely on a large group of part-time workers to help out, mostly on weekends and other busy periods.

The turbine airplanes used by these operations can cost more than $1 million each, and many big drop zones have several of these planes. Skydiving centers with turbine airplanes are often subject to special insurance restrictions that dictate pilot training and experience standards much higher than FAA minimums. Even the schools at large drop zones usually have several hundred thousand dollars invested in student and tandem parachute gear. Large drop zones that have so much money invested in equipment and staff tend to place a strong emphasis on safety.

SMALL DROP ZONES

Small drop zones are frequently started when a few skydivers get together at another DZ, and decide they want their own place to skydive. Sometimes these DZs develop because jumpers at a large drop zone want a smaller place to hang out, or because they think they can run a business better than the drop zone where they have been spending time. Sometimes a small DZ begins as a group effort, and sometimes it is a privately held business owned by just one jumper.

Starting a small drop zone requires an investment of less than $100,000, mostly used to buy a small Cessna 182 airplane, a few student rigs, a tandem rig or two, and a few simple training aids. The drop zone needs to rent space at an airport, and sometimes obtain a permit from the state government, an easy process in most cases. Most small DZs begin as weekend-only operations with the owners doing all the work, including teaching students. Often the people running the DZ and teaching skydiving have other jobs during the week.

Most of the people running small DZs really love skydiving, and will try to make their drop zone inviting to students. When you arrive at a small DZ you will often see just a few people hanging around, and you may feel like you are intruding on their private space. If you speak up and introduce yourself, you will probably find the skydivers are extra friendly, and happy you have stopped by. The first people you meet may even turn out to be your instructors, and within a few minutes you will probably become very comfortable with the group.

Small drop zones are sometimes managed as a communal activity. Almost all the experienced jumpers at these DZs have responsibilities ranging from packing student parachutes to pumping gas for the airplanes, or running a very informal manifest system. Small drop zones may have a full-time employee or two, but most rely on the support of the regular jumpers to handle critical chores, and many are not even open during the week.

Instructor Experience

The instructors at small drop zones may have the same ratings as the staff at bigger DZs, but since they don't handle the same student volume it is rare that they will have as much teaching experience. While you may find that some of the instructors at small drop zones lack the experience of their colleagues at big DZs, you will probably find that they are equally knowledgeable, attentive to your needs, and will take a very personal interest in helping you learn to skydive. Small drop zones usually concentrate their training on static line and tandem jumps, and sometimes have only one airplane that doesn't go much higher than about 10,000 feet.

Parties at small DZs tend to be more subdued than at big DZs, and boogies are often focused more on the celebration of a holiday. Since small drop zones are usually supported by groups of friends, you will often find them very inviting places to hang out.

Most small drop zones welcome spectators and encourage students to enjoy a picnic lunch on the airport grounds. You

Figure 9.4   Spectators are usually welcome at skydiving centers.

will probably find these drop zones lack the more elaborate facilities of big drop zones, but the jumpers who spend time at the DZ will make sure all the necessities are taken care of.

Small drop zones are often located in rural areas, and may be the only DZ for hundreds of miles. In many cases, the DZ will be the only business at an otherwise sleepy airport, and the jumpers will be welcome by the local community. In many small towns skydiving is considered a great use of the airport, and neighbors are happy to have you visiting their community.

Since the cost to open a small drop zone is relatively low, you may find some small DZs operating on a shoestring budget, and struggling to stay in business. Aircraft may be old and poorly maintained, and parachute equipment may be unsuitable second-hand gear discarded by another drop zone. You should be especially attentive to the condition of the equipment and knowledge of the staff at these drop zones. Of course, if a small DZ is well managed, as most are, you will be rewarded with a great first skydive, and may make some unexpected new friends.

MAKING IT ALL WORK

Drop zones can be busy places. Large DZs often have dozens of jumpers on staff, while small DZs often spread the workload among club members. Some drop-zone jobs pay well enough to provide full-time support, while others can offer a skydiver quick payment for a few minutes of work between jumps. Many skydivers who have regular full-time jobs supplement their income by helping out around the drop zone. There are some jobs that are especially well suited for students or inexperienced jumpers, and others that require years of training and experience. Some of the jobs you will find on a drop zone are discussed in this chapter along with the qualifications required.

Keep in mind that payment and job availability will vary greatly from one drop zone to another, and will differ around the country. In many cases the staff at a small drop zone will only be working part-time, and in some cases they may even be volunteers who are not paid at all for their work.

Manifest

A manifester works at the core of a drop zone. The manifest station is the critical link that holds the entire operation together, so the people working in this position are always under pressure. At a large DZ the manifester is frequently responsible for greeting new jumpers and making sure that all the required paperwork is filled out, dispatching airplanes, introducing jumpers to each other, collecting money, and coordinating the overall flow of activity throughout the drop zone.

A mainifester at a small drop zone may be the nonjumping husband or wife of a skydiver, or the owner of the operation. In some cases, small drop zones may use an honor system for manifest, with skydivers simply adding their names to a sheet of paper for each jump, and then settling their accounts at the end of the day.

A manifester at a large drop zone is often a full-time employee, with additional part-timers added on weekends. Some drop zones have separate manifesters for the school and the main operation. Manifesting is a great way for beginners to make some extra money to help with jump expenses, and since this position is at the core of the operation, manifesters get to know everybody and frequently enjoy high social status on the DZ. Once you get started with your skydiving training, you may have a chance to take on this important job.

Packer

If you are a student looking for a way to help pay for your skydiving, packing may be the job for you. Many skydivers consider packing parachutes drudgery. Often skydiving teams hire packers so they can spend more time training without the distraction of caring for their gear. Even jumpers who are not on teams often like to take a break and let somebody else pack their parachutes. Most skydiving schools will also need people to pack student parachutes and tandem rigs.

FAA regulations specify that parachutes must be packed by the person using them, a parachute rigger, or a person under the direct supervision of a rigger. Frequently, drop zones have a rigger train and supervise other people in the packing of their customers' parachutes. Sometimes this relationship is very formal, but sometimes it can be quite casual. In many cases, jumpers simply look around for somebody who is not doing anything and ask for help with a pack job.

Figure 9.5   Packing parachutes is not difficult.

Packing parachutes is not very difficult. In fact, you will probably learn the basics after just a few student jumps. Learning to pack well enough to work for other people takes a few days of practice, but many drop zones welcome your help, and gladly provide interested students with the needed training. A main pack job usually pays about $5.00, and a tandem pack job about $10.00. A quick packer can easily pack a rig in less than 10 minutes, and can earn several hundred dollars a day.

If you are interested in helping out with parachute packing, you should speak with an instructor and receive training from a rigger. Your instructor will be able to help you find a local rigger willing to supervise your work and help you meet the drop zone packing rules. Many students and beginning jumpers find that packing parachutes is a great way to make extra money and subsidize their training.

Rigger

A parachute rigger who holds an FAA certificate has completed training in packing and maintaining parachutes, and passed a written test, an oral test, and a practical exam. It is not especially hard to earn a rigger certificate, and you don't even need to be a skydiver, but learning the skills does take some time and effort.

It is very important for a drop zone to have several riggers on hand to comply with FAA regulations about packing main parachutes. Large drop zones often have many riggers available, but small drop zones sometimes don't have any. It is always a good idea to know who packed the parachute you will be jumping, and if the drop zone has a rigger employed to provide the needed supervision.

Parachute riggers are sometimes hired to work for a drop zone equipment dealership at a large DZ. Alternatively, they can be paid for each pack job they do. A typical parachute rigger will charge about $40.00 to $60.00 to inspect and pack a reserve parachute. Every main and reserve parachute must be inspected and packed no more than 120 days (about 4 months) prior to use, so riggers spend a great deal of their effort just inspecting reserve parachutes and keeping them legal for use. Parachute riggers who work full-time can make $30,000 per year or more, depending on how busy they are. Most parachute riggers work only part-time and use the limited income to supplement their skydiving costs.

Freefall Photographer

Freefall camera flyers are used by teams and schools to document skydives. The photographers at a large skydiving school usually have made more than a thousand jumps, and have lots of experience flying a camera. Camera flyers who shoot for teams may have much less experience. Getting started in camera work requires an investment of several thousand dollars in equipment. Flying a camera is more dangerous than regular skydiving, so you should be completely confident in your skydiving abilities before you try jumping with a camera. Many skydivers wait until they have made several hundred jumps before trying to jump with a video or still camera.

When you are ready to fly a camera you can often get friends or other jumpers to pay for your jumps in exchange for shooting them in freefall. After a bit of experience you will be able to start charging a few dollars per skydive, plus have your jump paid for. Successful camera flyers with skill and experience can then begin photographing students for the school, or develop their own style by shooting other experienced skydivers. Many freefall photographers really enjoy the unique creative opportunities that skydiving offers, and become experts in flying their own bodies to produce smooth, controlled, and well-composed photos or video. Some skydivers are so enthralled with combining freefall and photography that they become committed to flying a camera on every jump, even if they are not being paid for their efforts.

Camera flyers working for skydiving schools can make about $20.00 to $40.00 per jump, and if they work full-time at large drop zones, can earn more than $30,000 per year. Most camera flyers need to pay for all their own skydiving and video equipment, and do not have paid medical insurance, vacation time, or other benefits.

Coach

Skydiving coaches certified by the USPA or Skydive University assist with ground training and accompany upper-level students

Figure 9.6 Skydiving photography requires extensive skill and experience, as well as thousands of dollars worth of equipment.

in the air. This position requires more than 100 skydives and a specialized rating. Coaches are usually paid by the school based on how many jumps they make each day, or in some cases they are paid directly by the students. Coaches at big drop zones can make several hundred dollars a week. Coaches at small drop zones often make much less, or even work as volunteers.

Instructor

Skydiving instructors certified by the USPA provide students with training in AFF, static line, and tandem programs. At many skydiving schools, instructors are paid based on how many student jumps they make each day. A busy tandem instructor at a large drop zone can make a dozen jumps in one day, and earn more than $400.00. An AFF instructor usually spends more time teaching on the ground, and does not make as many jumps, so will generally make less for a day of work. Static line instructors are paid very little for each student with whom they work. A good instructor at a large drop zone with several ratings will often make AFF, static line, and tandem jumps, and can earn more than $1000 in a good week. Unfortunately, the work is weather-dependent, and very few instructors receive medical insurance, vacation days, or any other benefit plan.

Instructors who work in the northern part of the country in the summer, and then travel south for the winter, can stay busy all year, and can make an annual income of about $25,000 to $40,000. Instructors at small drop zones usually make far fewer jumps and earn less money. Most instructors work only part-time, and use the extra income to supplement their other jumping costs.

Some students recognize the work of a good instructor by offering a tip, often in the range of $10.00 to $20.00. Instructors understand that students are already paying quite a bit for their training and they do not expect a tip, but they do appreciate it when one is offered.

Pilot

Every skydiving center needs at least one pilot. You may find that at many small drop zones the pilot is a relative beginner who just barely meets the FAA minimum of 250 flight hours, while other small drop zones use pilots with thousands of hours of experience and many years of flying jumpers. The pilots at small drop zones usually make only a few dollars for each flight, and in many cases are doing the job simply because they love flying. Some pilots may even be working for free in order to build flight time and move to a better job. If you are a pilot and have an FAA commercial certificate, many small drop zones will be happy to work you into a flying position.

Pilots at large commercial centers usually need to meet much more stringent hiring standards. In almost every case, pilots flying turbine jump planes need to meet insurance company minimums that are far greater than FAA requirements. These pilots may also be trying to build experience, but almost all have at least 1000 flight hours before being placed in charge of a multiengine turbine airplane. Pilots flying at large drop zones, who are full-time employees, can expect to start earning about $20,000 per year.

Other Jobs

Skydiving centers are busy and dynamic places that require the services of all kinds of professionals. If you have an interest in skydiving, and have unique skills, you can often find a way to barter for jumps and training. Even if you lack special skills, most drop zones have a need for people to help in a variety of ways. Large drop zones need people to work in the office answering phones and helping answer student questions. Drop zones also need maintenance help, and some DZs need people to work in on-site restaurants or gear stores. Many of these jobs can be filled by people with little or no skydiving experience.

If you are looking for ways to subsidize your skydiving lessons, the staff may be able to find a few odds and ends that need to be handled. It is generally easier to find work on a drop zone once you have started training and had a chance to fit into the social scene. Skydivers understand that training is expensive, and most jumpers will do whatever they can to help an interested student complete the training program.

Chapter 10

MAKING YOUR DECISION

There are roughly 350 skydiving centers in the United States.[29] You may find it difficult to select the best one for your training, but with a little effort you should be able to find the one drop zone where you feel most comfortable. That selection should be based on your interests and desires, as well as the quality of the school itself. You will find that each drop zone is different, and although a few should be avoided, most are well managed and very professional.

There are many ways to choose a drop zone. Some people look in the Yellow Pages and then simply go to the drop zone with the biggest ad or lowest prices, while others do far more research, weigh the alternatives, and then select the skydiving school that seems just right for them. Picking a drop zone is really about making choices, and evaluating a vast array of options. Ultimately, your selection of a skydiving school will depend on what level of training you want, and what kind of consumer you are.

You may decide to learn at a small drop zone that has just one or two instructors, or you may want to make your first jump at a major center with dozens of instructors who train thousands of students each year. If you are sure you want to make only one jump in your life, you may be happy with a sky-diving school that offers nothing more than an exciting tandem ride. If you might want to make more jumps later, or are interested in seeing advanced skydivers in action, you may want to begin your training at a world-class school that can teach everything from the first jump to competitive formation skydiving, sky surfing, or freeflying. You may select a local drop zone where you can easily make one jump after work, or you may prefer to spend several weeks of vacation time at a top skydiving school far from home, and make dozens of jumps in a much more intensive program.

This book should have helped you to become an exceptionally well-informed consumer. At this point, you probably

Photo by Craig O'Brien

Figure 10.1 Large drop zones can provide training in many specialties such as sky surfing.

understand many of the training options available, and you may already know what you want to look for in a skydiving school. As you compare schools, keep in mind that your first jump can cost $150 to $300, and that's a lot of money. Many drop zones are competing for your business, so you should feel comfortable shopping around and making comparisons. Each time you check an ad or contact a drop zone, evaluate the school just as you would evaluate any other company that wants your money. Insist on a skydiving school that values your business and treats you with respect.

GETTING STARTED

You should begin your search by finding at least three local schools that will meet your minimum needs, and at least one school outside your area that is truly world-class. Next, check the web site for each school and call each of them on the phone. Compare their programs and make a decision. Then, visit your first choice and get a feel for that drop zone in person. Once you are armed with plenty of information, and actually see at least one drop zone in action, you will be in a great position to select the best school for your training.

Locating skydiving schools is easy. The United States Parachute Association has a web site at http://www.uspa.org/ that lists all the drop zones that belong to the organization. (See Figure 10.2.) The USPA site features basic information about each drop zone, including the types of training offered, airplanes that are used, distance to nearest major city, phone numbers, direct links to the official site for each drop zone, and much more. You can find other drop zones by looking in the Yellow Pages of your phone book under "Parachute Jumping" or "Skydiving Instruction." You can also use any major Internet search engine, such as Yahoo or Ask.com, and plug in

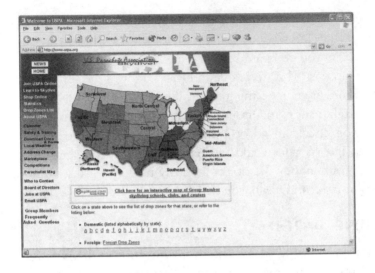

Figure 10.2 USPA org is a great place to locate a drop zone. (© *USPA*.)

the name of the closest big city, along with key words such as skydive, parachute, tandem, static line, or Accelerated Freefall. There are a few privately managed Internet pages that list drop zones, and most key-word searches based on "skydiving" or "parachute" will find those sites. Be cautious when considering drop zones listed on web sites other than uspa.org. Some web sites may appear to list affiliated and certified skydiving schools, but there may not be any actual oversight of the drop zones. Of course you can ask friends for recommendations too. When other people know you are interested in making a skydive they will usually offer their own first jump experiences, or put you in touch with people who have recently tried skydiving.

If you live in a metropolitan area it should be no problem to come up with at least three or four skydiving schools that are nearby. If you are in a rural area, there may only be one drop zone within easy driving distance. If you can't find more than a couple of local drop zones, select a few from other parts of the country to use for comparison. It's always best to compare several schools, rather than try to evaluate just one by itself.

CHECKING THE WEB

Once you have found a selection of drop zones, begin visiting their web sites. You will probably find that a drop zone with a professional-looking web site will also have a well-managed and professional training program. The ideal site will offer an overview of first-jump options and prices, a syllabus for a complete training program, a list of key staff and their qualifications, pictures of the facility, directions from a major city along with mileage and driving time, and phone numbers. A good site will be written in simple terms that are easy to understand, and will describe the programs offered in terms of the industry standards of tandem, Accelerated Freefall, and static line. If a school offers its own hybrid program, it should be detailed on the site, and contrasted with national standards so that you can easily compare it to the competition.

A good drop zone will have an active student training program, but it should also have a large group of experienced skydivers. The web site should be easy for students to understand, but it should also provide experienced jumpers with the information they need to evaluate the DZ. If the site has an area for experienced jumpers, check that out too. Although the information may not be of direct interest to students, it will help you to understand how the drop zone addresses its regular customers, and what you should expect when you visit. Most sites include a calendar of upcoming events, and this feature should give you an idea of the overall activity level of the operation.

Keep a list of all the drop zone web sites you visit, and write down their phone numbers. Remember, your goal is to

Figure 10.3 A good drop zone will appeal to both students and experienced skydivers.

select at least three local skydiving schools, and one distant school that you would like to investigate further.

USING THE PHONE

Once you have an overview of what each skydiving center offers, it will be time to call them on the phone. You should approach each call as an important interview and remember that you are the consumer, and must decide which school deserves your business. When you are talking with the schools on the phone, listen for a sense of respect. The person you are speaking with may be the owner of the school, a very experienced skydiver, or an office worker with just a few jumps. Keep in mind that a beginner may not be able to answer all of your questions, but will be in a great position to tell you about the program from a student's perspective. In any case, your questions should be answered in a meaningful way, and you should be comfortable with each response. If you don't understand an answer, ask follow-up questions until you have the information you need. If the person you are speaking to can't answer your questions, politely ask to speak with somebody who can. If the school brushes you off, or sends you back to the web site, be fore-warned, they probably don't value your business.

There is no such thing as a perfect drop zone. Each person will define the ideal skydiving school in his or her own way, so keep an open mind as you make your phone calls. Take notes, and if necessary, call back later for clarification.

Keep in mind that the school will almost certainly be trying to sell you on its own jump program. Listen carefully for promises that sound too good to be true, or anything else that just isn't believable. A reputable school will be proud of its training program and will focus its efforts on explaining the benefits of that program. Good schools will avoid discussing the competition in starkly negative terms, but they may offer a direct comparison if you ask. Be cautious of a school that spends too much time bad-mouthing other drop zones.

The best way to start a conversation is to ask who you are speaking with, and if that person is a skydiver. An active jumper should be able to quickly summarize his or her own experience and credentials. Let the person on the phone know you have never made a jump, then ask for an overview of the training options and costs. Most students ask similar questions, and a person experienced with phone work should be able to answer them quickly and easily with just a general prompt.

There are a few specific questions you can ask that will help you to evaluate the school. Answers to some of the questions are simple, while others will require the school to offer a bit more discussion. If you do not get a complete answer, you should ask direct follow-up questions until you are satisfied.

Questions to Ask

Is your drop zone a member of the United States Parachute Association (USPA)?   This is a key question, and you should listen carefully to the answer. There are a few quality drop zones that are not affiliated with the USPA, but

without this important affiliation you will have a difficult time
determining if the school meets any national standards. Some
schools that are not affiliated with the USPA will try to con-
vince you they are by saying that their instructors are members, or that they use USPA-trained instructors. Be extra
attentive to the answer you receive, and favor the drop zone
that is affiliated with this national organization.

**Are all of your instructors appropriately rated by the
USPA?**   This is another key question. Every instructor
should have a national certification specific to the program
being taught. Even drop zones that are not affiliated with the
USPA should use USPA-trained instructors. Some drop zones
may hedge a bit at the words "all" and "appropriately rated" in
your question. A clear answer to this question ensures that a
tandem instructor is actually rated to teach tandem, an
Accelerated Freefall (AFF) instructor is rated to teach AFF or
harness-hold training, and a static line instructor is rated for
that program.

Figure 10.4   The USPA offers a national
standard for skydiving centers. (© *USPA*.)

How many instructors do you have on staff?   A good school will have more than just one or two instructors. A larger staff will make it easier for you to find one instructor with whom you are really comfortable. The other advantage of a large staff is that instructors often learn great teaching tips from each other, so a big staff that interacts well with each other will generally improve the overall training level.

How much experience do your instructors have? This is a general question that can be answered easily, or it can be used as a more open-ended point of discussion. Obviously, an instructor with lots of experience is a good thing, but new instructors can also be advantageous. Beginning instructors who have just recently earned their ratings will have been trained in the latest teaching techniques, and will frequently be filled with enthusiasm. That new knowledge and enthusiasm will have a beneficial effect on the student program as a

Photo by Craig O'Brien

Figure 10.5   A large drop zone such as Perris Valley Skydiving Center, near Los Angeles, will have many different instructors available.

whole. A great staff will have both experienced instructors to serve as mentors, and well-trained new instructors and coaches. You should try to favor a drop zone with an active and vibrant staff of differing experience levels.

Who packs the student parachutes?   Most schools use regular jumpers to pack student parachutes, and provide an FAA-certificated rigger to supervise the packing. This is a common way to have the student gear packed, but many schools will be evasive about it. Listen carefully to the way this question is answered, and then follow up by asking how many riggers the school actually has on staff. A drop zone with several riggers available will generally be quick to identify maintenance problems, and will take good care of their equipment.

What experience and ratings do your pilots have? The FAA requires jump pilots to have at least a commercial pilot certificate and a minimum of 250 hours of flight time. More flight experience is always better, with about 1000 hours a reasonable amount of flight time for a beginning professional skydiving pilot. Some skydiving pilots have many thousands of hours and an Airline Transport Pilot certificate, a higher level of certification than a commercial pilot. Many people you speak with at the school will not know much about the certifications of their pilots, so you may receive an evasive answer. If this is a concern, you can ask the person on the phone to get the specific information you need and call you back.

How dangerous is skydiving?   This is a great open-ended question. Skydiving is risky, and there is really no perfect way to define the safety of this sport. You should expect an acknowledgment that injuries or even death can occur, and then anticipate a discussion of the drop zone's specific safety record. If you do not receive a firm answer about the specific safety record at that drop zone, ask a direct follow-up question about the local injury rate. Student fatalities are rare, but every active skydiving school has had at least a few student injuries, often just minor ankle or leg problems. The way the school addresses this concern should help you to gauge its credibility. Be very cautious of a school that tells you skydiving is absolutely safe, claims a perfect safety record, or quickly brushes off your concerns.

How long have you been in business?   A good school will often rely on word of mouth from satisfied students to attract business. A school that has been around for more than a couple of years will probably have a reasonable reputation, and a satisfied customer base.

How many students do you train in a year?   This question should help you to figure out if the drop zone is small or large. You can be sure a school that trains lots of students has developed an efficient means of teaching. A small drop zone with a limited student program is not necessarily a bad choice, but you should know about the volume of business when you make your decision.

How many students are trained at a time?   This is slightly different than the question about annual numbers. Big drop zones sometimes have huge classes that offer very limited personal attention. A good drop zone has small classes with a great instructor/student ratio. Class size is an important concern, and the drop zone should be able to provide a reasonable answer to this question. Be cautious if the school is evasive in addressing a question about class size.

A good follow-up is to ask how much time the instructors will actually spend with you. Some schools rush students along and allow very little time to meet and learn from an instructor. A few large schools use unrated teachers for ground training, and don't even let you meet your instructor until it is time to get on the airplane. You should use the answer to this question to help develop an idea of how much personal attention you will receive.

What kind of training do you offer?   A great drop zone will have an AFF program, or a hybrid program that includes AFF-style jumps conducted by USPA-rated AFF instructors. Even if you choose to make your first jump in the tandem or static line program, you will probably find that a drop zone offering an AFF program provides far better training than one without it. Likewise, a static line or tandem instructor who has earned the AFF rating will be better trained and tested than one who has not earned this elite rating. As you think about the answers you receive, keep in mind that many small drop zones do not do enough business to justify an AFF program, so you may not be able to find a school in a rural area that has this training available. Every school located near urban areas should have AFF training

Photo by Kaz Sheekey

Figure 10.6   A great skydiving center has AFF training available.

available, and you should be cautious about any that do not provide this level of training.

Does your school have a Skydive University program?

This is a tough question. If the school has this program, or has Skydive University coaches on staff, you should consider it a big plus, although you should not dismiss a school without Skydive University training. The Skydive University program is very rare in the United States, but it does offer great training, and should be considered a benefit. As with an Accelerated Freefall program, a Skydive University program will tend to elevate the level of training in other programs. Even if you have no interest in continued training, Skydive University will probably add quality to a drop zone at every level.

How old is your student equipment?

This is another good question that can generate an open-ended discussion. Skydiving gear wears out over time, and should be replaced regularly. New equipment is always being designed, and the latest gear will have the newest safety features. Equipment in use at even the best drop zones will vary from being brand new to several years old, so don't expect a very specific answer to this question. Instead, listen to how the question is handled, and try to gauge if the school has a replacement schedule for its gear. A replacement cycle of 3 to 4 years is pretty good, with some schools replacing their gear even more often.

Does every student rig have both an automatic activation device (AAD) and a reserve static line (RSL)?

These critical safety features are mandated by USPA Basic Safety Requirements (BSRs), so all member drop zones should say yes without hesitation. Any deviation from this answer is a great reason to avoid the school completely.

Will I be given an altimeter?

Again, USPA BSRs require that every student have an altimeter, so all USPA-affiliated drop zones should respond with an immediate yes. A few schools do not provide their tandem and static line students with altimeters. The lack of an altimeter should be a matter of concern because it is required by the BSRs. A USPA drop zone that violates this simple rule may be breaking other, less obvious rules, too.

How many people can go up at once?   The answer to this question should include the kind of airplane the drop zone is using, and that will help you to gauge the size of the operation. If you are making your skydive with a group of friends, a small airplane will probably be able to take only one or two of you at a time. A large airplane will be able to accommodate more students at once. Keep in mind that big airplanes are very expensive. A drop zone flying a Twin Otter, King Air, CASA 212, or similar large turbine airplane will consider it a valuable asset, and will probably have a solid injury-prevention program to protect that investment.

From what altitude do students jump?   Small drop zones using a typical Cessna 182 sometimes go up to only about 9000 or 10,000 feet for AFF and tandem students. Larger drop zones with turbine airplanes frequently jump from as high as 14,000 feet. Some big drop zones limit their student jumps to just 9000 feet to save money, while others offer a low price on the phone for limited altitude, and then try to sell a more expensive jump from higher altitude when you arrive. A typical tandem jump from 9000 feet will offer only about 25 seconds of freefall, while a jump from 14,000 feet will offer closer to 50 seconds. When evaluating drop zones, make sure the altitude offered is comparable, and favor the drop zone that offers higher jumps. Static line jumps are usually made from 3000 to 3500 feet. Again, favor the drop zone that offers a higher altitude, and avoid one that has students jumping from any altitude lower than 3000 feet.

When is the school open?   This question will help you determine if the school is a full-time operation, or a weekend-only drop zone. This will also help you figure out how easy it will be to schedule your first skydive, and future training, if you decide to make more jumps.

How much does it cost?   This is often the most important question for many people. While cost should always be considered, it should not be the deciding factor when choosing between schools. Think about the answers to all the other questions first, and then evaluate the cost. Remember, the most expensive program is not necessarily the best, nor is the least expensive school always a good value.

COMPARING THE ANSWERS

When you have finished calling all the drop zones on your list, you should compile the answers you have received and compare them. Keep in mind that a good salesperson can use attitude and fancy language to sell just about anything, so begin your comparison by looking at the answers to the objective questions. Compare the kind of airplanes, types of programs, instructor ratings and certifications, jump altitude, location, AAD and RSL use, altimeter use, class size, and USPA affiliation. Then, add in the general feeling you had talking with the people who answered the phone.

After you have checked the web sites and spoken directly with people at the drop zones, you will be ready to make a decision and select one school for your training. Ideally, you should visit the drop zone before you commit to make a jump. When you make that visit, compare what you see with what you were told on the phone. Check the general condition of the buildings and property, and the appearance of the airplanes. Get a feel for the activity level, and the mix between first-jump students, satisfied repeat students, and experienced jumpers. Watch a few of the instructors as they teach, and assess their "people skills." Talk with the instructors and ask them some of the same questions you asked on the phone. Listen for consistent answers, and try to gauge your comfort level with the drop zone.

A good drop zone will welcome your visit and take advantage of the opportunity to show you around. The staff should be willing to let you watch some parts of the training, and should be able to show you video of actual student jumps. Most of the staff will probably be busy and will not have much time to accompany you, but they should at least make you feel welcome, and provide some access to their training environment.

MAKING THE DECISION

If you are happy with the quality of the drop zone and the staff, go ahead and make an appointment for your first jump train-

Figure 10.7   Visit several drop zones before making your decision.

ing. If you are planning to make a tandem skydive, the school may be able to accommodate your request right away. Static line and Accelerated Freefall classes usually start early in the morning, and frequently fill up fast, so you will probably need to schedule an appointment and return for that training.

When you are ready to actually start your first-jump training, plan to arrive early. Bring a photo I.D., comfortable clothing, and sneakers or light boots. Your instructors should begin the training with a brief introduction that includes their specific experience and the ratings they hold. If your instructors do not provide this information, you should ask for it. Every USPA-rated instructor is issued a membership card that includes his or her ratings. (See Figure 10.8.) If you have any concerns about your instructor's experience, you should feel free to ask for this proof. Students generally just take an instructor's word that he or she is appropriately rated, but you always have the option of requesting written confirmation if you have any doubt. Some instructors may not have their membership cards handy, but a USPA-affiliated school can easily access a private part of the USPA web site and provide you with an official printout showing each instructor's ratings. (See Figure 10.8.)

You should have an opportunity to see the gear you will be using as you prepare for your skydive. A good instructor will show

Figure 10.8   The school should be willing to show you proof of instructor certifications. (© *USPA*.)

you how it works and point out all the important safety features. If your instructor doesn't do this, you should ask for a detailed explanation of the equipment.

Sometimes things may seem a bit rushed. If you need more time to understand the training, or want additional help, simply ask. Most instructors will be happy to slow down and help you become comfortable before you make a jump. Always keep in mind that you are the consumer, and your safety is critical. If you are unhappy with the quality of the training or the equipment, you have the option to say no. You should never feel pressured into making a skydive or doing anything in the sport that you are not comfortable with.

You will probably find that most skydiving schools offer excellent training with great staff, and the research you do ahead of time should help you to become comfortable with your decisions. Skydiving is really fun, and you will almost certainly have a blast on your first jump. It is always nice to share the experience with other people, so do your research, grab a few friends, and then head out to the drop zone.

FREQUENTLY ASKED QUESTIONS AND ANSWERS

Skydiving schools have become accustomed to answering questions from prospective students. Many of the most common questions asked by beginners are listed below as a standard FAQ (Frequently Asked Questions), along with a chapter number so you can easily look up more information.

Q. How do I find a skydiving school?

Many skydiving schools are listed in the Yellow Pages of your phone book. You can also check on the web at http://www.uspa.org, or use any of the available search engines such as Yahoo. If you have friends who have jumped, you should ask for their recommendations. (Chapter 10)

Q. What is a tandem jump?

When you make a tandem jump you wear a harness that is attached to your instructor's. Your instructor has a main and a reserve parachute. After a freefall of 20 to 50 seconds, either you or your instructor pulls a ripcord. You remain connected and fly a parachute built for two. Tandem skydives are generally made from about 8000 to 14,000 feet and involve both a freefall and a parachute descent. (Chapter 2)

Q. What is a static line jump?

When you make a static line jump your parachute is attached to the airplane via a static line, and it opens automatically about 8 to 12 feet after you jump. Static line jumps are usually made from about 3000 to 3500 feet. (Chapter 2)

Q. What is Accelerated Freefall?

When you make an Accelerated Freefall jump you wear your own parachute rig. One or two instructors leave the airplane with you between 9000 and 14,000 feet. The instructors hold onto you throughout the freefall. You pull your own ripcord after about 30 to 50 seconds. Once you pull the ripcord your instructors fall away, and you fly your own parachute back to the drop zone. (Chapter 2)

Q. Do I need to make a tandem jump first?

Some skydiving schools require a tandem jump before they will allow you to make an Accelerated Freefall or static line jump. Other schools allow you to make a static line or Accelerated Freefall without first making a tandem jump. Many instructors believe that making a tandem jump first helps reduce landing injuries, and makes it easier for you to master advanced skills. (Chapter 2)

Q. Can I jump when I'm drunk?

No. (Chapter 9)

Q. Can I bring people with me just to watch?

Sure. Many skydiving centers have special areas for spectators. Your friends and family will almost certainly enjoy watching you make your first jump. (Chapter 9)

Q. My ears pop in airplanes. Is that going to happen when I jump?

Some people have trouble clearing their ears after a skydive. If you think this might be a problem, you should speak with your instructor and he or she will offer ways to help you equalize pressure once the parachute opens. (Chapter 1)

Q. What if I get sick?

Very few people get sick when they make a skydive. Skydiving is not as likely to cause nausea as many amusement park rides. If you think this might be a problem, speak with your instructor. (Chapter 1)

Q. What if I black out?

This can happen on a tandem skydive, but it is very rare. Your instructor is trained to handle the situation, and it should not be a concern. (Chapter 1)

Q. What if I'm too afraid to jump?

You don't have to jump. Most people are apprehensive, and some become very scared just before leaving the airplane. This reaction is normal and generally goes away several seconds after leaving the plane. There are some easy ways to help control your stress, and your instructor should be able to help. If you really don't want to jump you should tell your instructor, and you will ride the plane down together. No one should force you to make a skydive. (Chapter 4)

Q. I'm really scared of heights. Is that a problem?

Not usually. You will be making your jump from so far above the ground that you will probably not notice this normal fear. Skydiving rarely bothers people with fear of heights. Many expert skydivers were once afraid of heights, but used skydiving as a way to conquer or minimize the fear. (Chapter 4)

Q. How long is the freefall?

It varies depending on the altitude you jump from and how high you open the parachute. A typical freefall for a tandem or Accelerated Freefall jump will be 30 to 50 seconds. There is no freefall on a static line jump. (Chapter 2)

Q. How long is the parachute ride?

It depends how high you pull the ripcord. A typical student or tandem parachute will descend at about 1000 feet per minute. Most tandem and Accelerated Freefall students open between 4000 and 6000 feet, so the parachute flight should last 4 to 6 minutes. Static line students usually leave the plane at about 3000 feet and the parachute opens immediately, so their parachute flight should last about 3 minutes. (Chapter 2)

Q. How hard is the landing?

Landings are usually very comfortable, but sometimes they can be hard. Your instructor will train you to deal with landings. (Chapter 5)

Q. How fast do skydivers fall?

Skydivers usually freefall at about 120 mph, but can go as slow as 40 mph, or descend at more than 300 mph. (Chapters 1 and 8)

Q. How long does the training take?

Training for a tandem jump should take about 30 minutes. Training for the first Accelerated Freefall and static line jumps should take 5 to 6 hours. Training for subsequent Accelerated Freefall and static line jumps should take about 45 minutes. (Chapter 2)

Q. How much does it cost?

Prices vary around the country, and even between drop zones in the same area. A first tandem skydive is usually about $150 to $250. A first static line jump is usually about $125 to $175. A first Accelerated Freefall jump is usually about $250 to $325. (Chapter 2)

Q. Are discounts available?

Some schools offer college and group discounts, or weekday discounts. You should ask the school about this directly. (Chapter 1)

Q. Should I tip my instructor?

Instructors always appreciate a tip, but it is not necessary. (Chapter 9)

Q. What if the parachute doesn't open?

Every skydiver has a main and a reserve parachute. If you are making a tandem skydive, your instructor will handle any emergency. If you are making an Accelerated Freefall or static line jump, you will be trained to deal with malfunctions. In every case, student parachutes have many safety features, almost always including a device that can automatically open the reserve in an emergency. (Chapters 1 and 5)

Q. Is skydiving dangerous?

Skydiving is a high-risk activity with a potential for injury or death. If you take appropriate precautions you can reduce the risk and minimize the hazards. Great training and equipment, coupled with a safety-conscious approach, are the keys to making your jumps safe. (Chapter 7)

Q. How hard is the parachute opening?

Most parachute openings are very soft. Some openings can be hard, but it is rarely a problem. (Chapter 5)

Q. What are parachutes made of?

Modern parachutes are made of nylon. (Chapter 5)

Q. Are parachutes still round?

Almost all civilian schools use rectangular main parachutes. Most schools also use rectangular reserves, although a few still use round parachutes as reserves. (Chapter 5)

Q. Who certifies instructors?

No government agencies certify instructors. Most instructors have received training and have been certified by the United States Parachute Association (USPA), a membership group representing skydivers. You may find some instructors who are not certified at all. (Chapter 3)

Q. Are skydiving schools licensed?

About 80 percent of the skydiving schools in the United States are affiliated with the United States Parachute Association. These schools have agreed to follow a basic set of rules. Schools that are not affiliated with the USPA are usually not regulated at all. No federal license is required to operate a skydiving school. (Chapter 3)

Q. What is Skydive University?

Skydive University is a private company that conducts its own training programs, and certifies coaches and instructors. Skydive University coaches have met or exceeded the minimum instructional rating requirements of the United States Parachute Association. (Chapter 3)

Q. Can I request a male or a female instructor?

Most schools will try to accommodate special requests, but it is sometimes difficult to provide an instructor based on sex. Instructors are well trained and should treat you with respect and dignity, regardless of their sex. (Chapter 4)

Q. How many jumps can I make in a day?

Student jumps can be mentally challenging, so you probably shouldn't make more than two or three jumps in a day. (Chapter 2)

Q. How many people go up at once?

It depends. Some schools use airplanes that carry only 4 jumpers at a time, and some use airplanes that can carry 20 to 30 people. Most drop zones with big airplanes will try to keep groups of friends together whenever possible. (Chapter 6)

Q. Will I have to pull the ripcord?

If you make a tandem skydive your instructor may give you a ripcord and teach you when to pull, or he may simply do it himself. If you are making a static line jump, your main parachute will open automatically. If you are making an AFF jump, you will be responsible for pulling your own main parachute ripcord. (Chapter 2)

Q. How will I know when it's time to pull?

Your instructor should give you an altimeter, a sensitive device that will display your altitude. It is as easy to read as a wristwatch. If you are making a tandem skydive, your instructor can also shout in your ear when it is time to pull. (Chapter 2)

Q. What happens if I don't pull on a tandem skydive?

If you are making a tandem skydive and don't pull the ripcord yourself, your instructor will pull it for you. (Chapter 2)

Q. Who will pack my parachute?

Main parachutes must be packed by an FAA-certificated parachute rigger, the person making the jump, or a person under the direct supervision of a rigger. Most student main parachutes are packed by well-trained people under the supervision of a rigger. Your reserve parachute will be packed by a rigger. (Chapters 3 and 9)

Q. Do you always use two parachutes?

Yes. The FAA requires every skydiver to have a main and a reserve parachute. (Chapters 3 and 5)

Q. Does a skydiver really go up when he pulls?

No. When watching a video it sometimes looks like a jumper goes up when he pulls. This is an optical illusion created when a freefalling photographer continues freefall after the subject of the video has opened his or her own parachute. (Chapter 1)

Q. Do skydivers say "Geronimo"?

Not usually. (Chapter 4)

Q. Can we talk to each other in freefall?

Only in Hollywood. On a tandem skydive your instructor will be able to yell single words in your ear, but that is about it. On

Accelerated Freefall jumps, you will be traveling too fast, and there is far too much wind noise for real conversation. Tandem students are usually able to talk with their instructors once the parachute has opened. (Chapter 2)

Q. Can I jump in a cloud?

No. Skydivers avoid jumping in clouds because they can't see each other, or see airplanes that may be in the cloud. Sometimes a skydiver will jump through a small cloud, but that is very unusual. (Chapter 1 and Chapter 3)

Q. What if it is too windy?

Your instructors shouldn't let you jump in high winds. Accelerated Freefall and static line students shouldn't jump in winds greater than about 14 mph. Tandem instructors will decide if it is too windy for you based on several factors, but will usually not jump if the wind is much stronger than about 20 to 25 mph. (Chapter 1)

Q. I have a physical handicap. Is that a problem?

Not usually. There are blind and deaf skydivers, and even paraplegic and quadriplegic people have made tandem skydives. You should discuss any handicap with the school in advance. A good skydiving school will be able to help you obtain a booklet titled *Skydiving with Wheelchair Dependent Persons* that is produced by a company called The Uninsured Relative Workshop. (Chapter 1)

Q. I wear glasses. Is that a problem?

No. Skydiving schools have special goggles that will fit over your glasses, and some schools have straps to hold your glasses for extra security. Contact lenses are not a problem either. If you wear contacts, you should tell your instructor so you receive a tight-fitting pair of goggles. (Chapter 5)

Q. Is there an age limit?

Most skydiving centers have established a minimum age of 18, although a few drop zones may be willing to accommodate younger jumpers. There is no upper age limit. (Chapter 1)

Q. Is there a weight limit?

Parachute equipment is certified with a maximum weight limit. The specific limits vary, but many schools establish their own limit of about 220 pounds. Some schools will accommodate heavier students but may charge extra. If you think weight might be a problem, you should discuss it in advance with the school. (Chapter 1)

Q. How much does parachuting equipment weigh?

Student parachute rigs used for Accelerated Freefall and static line jumps usually weigh about 30 to 35 pounds. Tandem rigs weigh about 50 pounds. Your instructor will carry the tandem rig. (Chapter 5)

Q. Does my health and life insurance cover skydiving accidents?

It may. Some policies have exclusions that limit coverage in aviation or adventure sports. You should check with your insurance provider. (Chapter 1)

Q. Do I need to bring anything special with me?

Not really. You should have photo identification, boots or sneakers, and appropriate clothes for the weather, but that is just about all you will need. (Chapter 1)

Q. What should I wear?

You should wear comfortable clothes for the weather conditions. You should also wear sneakers or light hiking boots. Sandals are not a good idea for student jumpers. It is usually colder in the sky, so if it is a chilly day you should bring light gloves and a sweater or sweatshirt. (Chapter 1)

Q. Can I bring a camera?

Yes, but you will not be able to take it with you in freefall. Many students like to have a camera to shoot photos on the ground before and after the skydive. Most schools can arrange for a professional freefall photographer to accompany you on the skydive. These photographers will often take still pictures, and even be able to make a video of your jump. They usually skydive with you and wear the cameras attached to their helmets. (Chapter 9)

Q. Should I avoid eating before a jump?

No. It often helps to have something light in your stomach, such as cookies, popcorn, or a sandwich. Students should avoid eating greasy or spicy food before jumping. (Chapter 1)

Q. Can I breathe in freefall?

Yes. Some students have a slight problem breathing, but it is easy to deal with. Your instructor can offer some tips to help you. (Chapter 1)

Chapter 1

1. USPA web site, May 2002.

2. USPA web site, May 2002.

3. The British Parachute Association (BPA) medical statement notes that oxygen levels are reduced by 40 percent at 15,500 feet. The medical statement also notes that tachycardia of 120–160 beats per minute (bpm) is common in experienced parachutists, and 200 bpm is not unusual in novices. Tachycardia may be present at the same time as relative hypoxia.

Chapter 2

4. Recognized by the adoption of a rewritten 14 CFR, Part 105. The final rule change was published in the *Federal Register* on May 9, 2001, and became effective July 9, 2001.

Chapter 3

5. FAA Notice of Proposed Rulemaking, April 13, 1999.

6. Affiliation and membership levels provided by USPA, May 2002.

7. A USPA "A" license requires a minimum of 25 jumps, plus demonstration of specific skills. The required number of jumps was increased by USPA in September 2003.

8. A USPA "D" license requires a minimum of 500 jumps, plus demonstration of specific skills. The required number of jumps was increased by USPA in September 2003.

9. The minimum number of jumps required for each instructional rating is based on USPA license standards in effect in 2002. Licensing standards were increased effective September 2003. The minimum number of jumps required for an instructional rating did not change, but may be slightly modified in the future.

Chapter 5

10. Cypres Users Guide.

Chapter 7

11. USPA Report, May 2002.

12. USPA Report, May 2002.

13. USPA Report, May 2002.

14. 2001 data from *Parachutist* magazine, April 2002.

15. USPA issues licenses defined by jumper experience level. The licenses available are A, B, C, and D; and they require 25, 50, 200, and 500 jumps, respectively. Minimum required jump numbers for new licenses were increased by USPA in September 2003 to conform to international standards.

16. USPA web site, May 2002, and anecdotal reports.

17. NPRM, *Federal Register*, April 13, 1999.

18. Data provided by the Uninsured Relative Workshop, May 2002.

• 19. *Injury Facts*, 2001 edition, page 13.

20. *Injury Facts*, 2001 edition, page 16.

21. *Injury Facts*, 2001 edition, page 123.

22. *Injury Facts*, 2001 edition, page 124.

23. *Injury Facts*, 2001 edition, page 16.

24. *Injury Facts*, 2001 edition, page 122.

25. *Injury Facts*, 2001 edition, page 8, reported as 15.6 per 100,000 population.

26. Annual data provided by USPA, May 2002.

27. Reported by The Uninsured Relative Workshop, May 2002.

Chapter 8

28. USBA report, May 2002.

Chapter 10

29. Based on 284 USPA affiliated drop zones in 2001, with an estimated 80 percent of all domestic drop zones belonging to the organization, according to figures provided by the USPA in May 2002.

Glossary

AAD See **automatic activation device.**

AFF (Accelerated Freefall) A training program administered by the United States Parachute Association that features one or two instructors accompanying a student in freefall.

AFP (advanced freefall progression, advanced freefall program) One of several hybrid programs that have been developed by individual skydiving schools, often featuring one or more tandem jumps followed by harness-hold jumps.

AGS (advanced ground school) A ground school in a hybrid program, usually conducted between tandem jumps and harness-hold jumps. Often called a comprehensive ground school.

airgasim A good time in freefall (slang).

altimeter A device for measuring and displaying altitude.

arch A basic body position used for stable freefall.

ASI (advanced skydiving instruction) One of several hybrid training programs developed by individual skydiving schools, often featuring one or more tandem jumps followed by harness-hold jumps.

ASP (accelerated skydiving program) One of several hybrid training programs developed by individual skydiving schools, often featuring one or more tandem jumps followed by harness-hold jumps.

Astra A brand of automatic activation device manufactured by the FXC Corporation.

ATP (advanced training program) One of several hybrid training programs developed by individual skydiving schools, often featuring one or more tandem jumps followed by harness-hold jumps.

automatic activation device (AAD)   A mechanical or electronic device that senses altitude and freefall speed and activates a parachute. An AAD is usually fitted to a reserve parachute, but it can be used with a main parachute.

BASE jumping   Jumping from fixed objects such as a Building, Antenna, Span (Bridge), and Earth (Cliff).

Basic Safety Requirements (BSRs)   Minimum safety standards published by the United States Parachute Association.

beer rules   Informal rules at many drop zones that require a skydiver to purchase and share beer following significant accomplishments and milestones, or applied as a penalty when a rule is broken.

BOC (bottom of container)   A deployment system that features a pilot chute stored in a pocket on the bottom of the container. To deploy the main parachute a skydiver pulls the pilot chute out of the pocket and throws it into the air.

boogie   A big skydiving party, usually featuring increased jump activity during the day and a traditional party at night. Pronounced like a disco dance.

bounce   A term used when a skydiver dies or lands hard enough to cause a serious injury (slang).

brain lock   When a jumper forgets the maneuvers that were planned for freefall (slang).

breakaway handle   A handle used to release the main parachute from the harness.

bridle   A nylon line used to connect a pilot chute to the top of the parachute. The pin that holds the container closed is usually attached to the bridle.

BSR   See **Basic Safety Requirements.**

canopy   The fabric and suspension lines of a parachute.

canopy relative work (CRW)   Flying parachutes together to make canopy stacks or formations.

CGS (comprehensive ground school) An extended ground training course usually included in a hybrid training program. This course is generally taken after a series of tandem jumps, but before harness-hold jumps.

chamber test A test of an automatic activation device that involves placing the unit into an airtight chamber that will allow the pressure to be rapidly changed so the functions of the unit can be checked. Most automatic activation devices should be chamber-tested according to the manufacturer's maintenance schedule.

clear and pull A skydive in which the jumper leaves the plane and opens his or her own parachute immediately, with a minimum freefall. It is usually incorporated as part of static line and Accelerated Freefall programs.

COA (circle of awareness) A quick freefall performance check used by an Accelerated Freefall student. It involves checking the ground for a heading, checking an altimeter, and then looking to an instructor for signals or corrections.

Coach A skydiver authorized to teach limited skills. Coaches are usually certified by either the United States Parachute Association or Skydive University. Advanced training of specialized skills is often conducted by jumpers who have a great deal of experience in the specialty, or advanced training in sports instruction, but no formal rating.

creeper A small wooden or fiberglass platform on wheels, used to practice formation skydiving positions. A creeper is usually just large enough to support the torso of a jumper in an arch position. It is used to assist in moving along the ground as horizontal formations are practiced.

CRW See canopy relative work.

cutaway handle A handle that is used to release the main parachute risers from the harness. Usually used in an emergency when the main container has opened, but the parachute is not flying properly. Also referred to as a breakaway handle.

Cypres   A common brand of automatic activation device.

debrief   A discussion about a jump after it has occurred. A jump is usually debriefed shortly after landing, so that mistakes and successes can be identified.

demo jumper   A skydiver who makes public demonstration jumps away from the drop zone.

direct bag   A type of static line system that connects the deployment bag to the airplane. As the jumper leaves, the lines are deployed and the parachute is extracted from the bag. The static line and bag remain attached to the airplane. See also **pilot chute assist.**

dirt dive   A practice skydive conducted while standing on the ground. A dirt dive allows all the jumpers involved to memorize their intended movements, and to coordinate them with the other skydivers who will be on the jump.

dope rope   Static line (slang).

drop zone (DZ)   A field where skydiving takes place.

drogue   A small round drag device used to slow a tandem skydiving pair to about 120 mph. A tandem instructor usually throws a drogue shortly after leaving the airplane to improve stability and prevent acceleration to speeds that are higher than single-person freefall speed.

Dual Hawk   A brand of tandem rig manufactured by Strong Enterprises.

DZ   See **drop zone.**

Eclipse   A brand of tandem rig originally manufactured by Stunts Adventure Equipment but no longer being produced. Some drop zones still use this equipment.

elliptical   A ram-air or square-style parachute with tapered ends. An elliptical parachute offers greater performance than a conventional square parachute, but is generally more demanding to fly and can present more radical malfunctions.

eight-way   An eight-person freefall formation. A standard and popular size of formation used in competition.

exhibition jumper   See **demo jumper.**

FAA (Federal Aviation Administration)   The federal government agency responsible for regulating aviation and skydiving in the United States.

FARs (Federal Aviation Regulations)   The rules of the Federal Aviation Administration.

FJC (first jump course)   A comprehensive ground school required before making an Accelerated Freefall or static line jump. It usually takes 4 to 6 hours.

floater bar   A hand railing installed on the inside or outside of an airplane door for skydivers to hold onto while they prepare for exit.

four-way   A four-person freefall formation. A standard and popular size of formation used in competition.

formation skydiving   When two or more jumpers get together in freefall to build formations. Also known as relative work or RW.

frap hat   A soft leather skydiving helmet with small ridges to provide protection. It resembles an old-style football helmet. Tandem students often use frap hats.

freefly   Flying in many dimensions on one jump. Freefly skydivers like to fly on their bellies, on their heads, standing up, in a seated position, or any combination.

FS   See **formation skydiving.**

function   When a parachute does not work. Shortened from malfunction (slang).

FXC   An equipment company known for its automatic activation devices. FXC manufactures the electronic Astra and the mechanical FXC 12000.

ground heading   The direction a jumper is facing relative to a ground reference.

hand-deploy pilot chute   A pilot chute that is thrown into the air to begin the opening of a main parachute. Usually contained in a pocket of the parachute container.

hard arch A form of the basic arch position frequently used for beginning static line jumps.

hard deck A key altitude, usually a pull altitude or an emergency response altitude. A jumper should not go below a defined hard deck without taking a specific action related to the opening of either the main or reserve parachute.

heading A visual reference in freefall or when flying a parachute.

hybrid training Any skydiving training program that includes jumps from more than one of the standard disciplines of tandem, static line, instructor-assisted deployment, or Accelerated Freefall.

IAD (instructor-assisted deployment) A training method similar to static line. The student leaves the airplane, usually between 3000 to 3500 feet, and the instructor throws a hand-deployed pilot chute into the air to open the main parachute immediately.

IAF (instructor-assisted freefall) One of several hybrid programs that have been developed by individual skydiving schools, often featuring one or more tandem jumps followed by Accelerated Freefall-style jumps.

instructor A skydiver who teaches students. Most instructors have received training and have been certified by the United States Parachute Association.

ISP (integrated student program) A formal training program established by the United States Parachute Association that provides a student with the instruction needed to qualify for a license.

jump run The part of an airplane flight used for jumping. An airplane will climb to the designated jump altitude, then be aligned with the airport. Jump run begins when the airplane is established at the correct altitude, heading, and airspeed.

lazy w   A freefall position based on the arch, but using a modified and relaxed arm position.

license   A certification issued by the United States Parachute Association. A beginner's license is issued after about 25 jumps.

lift   One of the four aerodynamic forces. Lift is the force that allows a rectangular parachute to descend slowly and fly like an airplane wing.

log book   A book used to record details of a skydive A log book is required for advancement and license certification.

lurk   (1) To wait patiently for an available position or opportunity (slang). When a skydiving school is completely booked, a student may often wait and hope that a position on the schedule opens up. (2) A jumper in freefall who is watching a formation and waiting to become part of the group (slang).

malfunction   When a parachute does not work properly. Malfunctions can be total, meaning the parachute did not open at all, or partial, meaning the parachute opened, but did not provide an adequate reduction in rate of descent.

manifest   (1) The place where jumpers sign up for a jump. (2) The list of people on a jump.

meat hauler   A tandem instructor (slang).

parachute   A device to reduce descent rate so a skydiver can land without injury. A parachute is usually made of nylon. Some parachutes are round. Most parachutes used for civilian skydiving are rectangular.

parachutist in command   The official FAA term for a tandem instructor.

part 105   The FAA regulations that specifically cover skydiving. Part 105 is contained in the Code of Federal Regulations (14 CFR).

passenger parachutist   The official FAA term for a tandem student.

pilot chute A small round parachute, usually about 30 to 36 inches in diameter, that is used to open the parachute. Main parachutes are usually opened using a pilot chute pulled from a pocket and thrown into the air. Reserve parachutes are usually opened by a spring-loaded pilot chute that is contained inside the parachute pack, and released by pulling a ripcord. Some students use main parachutes opened by a spring-loaded pilot chute and ripcord assembly.

pilot chute assist A type of static line system that uses a static line connected to the airplane, and then to the top of the pilot chute. The pilot chute extracts a bag that holds the parachute and deployment lines. The static line remains attached to the airplane; the pilot chute and bag remain attached to the parachute. See also **direct bag.**

piston engine A type of engine usually found on small airplanes such as the Cessna 182.

PRCP (practice ripcord pull) A training exercise that requires a student to go through the motions of touching a ripcord without actually opening the parachute. A PRCP is frequently used on early jumps to build a memory pattern for the ripcord pull. Usually identical to a PRCT.

PRCT (practice ripcord touch) A training exercise that requires a student to go through the motions of touching a ripcord without actually opening the parachute. A PRCT is frequently used on early jumps to build a memory pattern for the ripcord pull. Usually identical to a PRCP.

prop blast The wind created by the propeller while holding onto the outside of an airplane, and immediately after exit.

pull-up cord A small ribbon of material about 16 to 20 inches long, used to help a packer close a parachute rig.

Racer A brand of tandem rig manufactured by The Jump Shack.

ram air A square-style parachute with an open front end and a closed back end. The parachute maintains shape when air is forced into openings on the front, and is trapped by the closed tail area.

relative work (RW) Building formations in the air, usually while in a belly-to-earth position. Also called formation skydiving.

reserve static line (RSL) A lanyard that connects the risers of the main parachute to the reserve ripcord or ripcord cable. An RSL is a backup device designed to open the reserve if the main parachute is cutaway. It is a backup only, and should never be relied upon.

rig A complete parachute system, including a harness/container, main parachute, and reserve parachute.

rigger A person who packs parachutes and has a certification from the FAA.

ripcord A device used to open a parachute. Pulling a ripcord opens the parachute container and allows a pilot chute with a built-in spring to launch into the air and begin the deployment process. All reserve parachutes used for civilian skydiving are activated with ripcords. Some student main parachutes are operated with ripcords, while others use a hand-deployed pilot chute.

RSL See **reserve static line**.

RW See **relative work**.

Safety and Training Advisor A senior skydiver who acts as a volunteer representative of the United States Parachute Association on a specific drop zone. Also known as S&TA.

sensory overload A brief period shortly after exit when a student skydiver is so overwhelmed by the experience that awareness is reduced. Memory of the first few seconds of the jump will usually be limited by sensory overload.

Sigma A brand of tandem rig manufactured by The Uninsured Relative Workshop.

SIM (Skydivers Information Manual) A book published by the United States Parachute Association that contains rules, regulations, and procedures that have been recognized as a reasonable industry standard by most drop zones and skydivers in the United States.

sitfly Flying in a seated position.

skydive A parachute jump, usually involving a freefall.

skydiver A person who makes skydives.

Skydive University A private company that certifies some skydiving coaches, and offers advanced training programs.

sky surfing Flying with a snowboard-like device attached to your feet.

S/L See **static line.**

snivel A parachute that is opening slowly (slang).

SOS (1) single operation system: A type of parachute system that uses one handle to both cut away the main parachute and activate the reserve. Used in some training programs, but never used by experienced jumpers. (2) Skydivers Over Sixty: A group of skydivers who are at least 60 years of age, and have made at least one jump.

spot (1) The place over the ground where skydivers should leave the plane to ensure they will be able to fly their parachutes back to the intended landing area. (2) The process of guiding the airplane to the correct location for a jump.

spring-loaded pilot chute A pilot chute with a built-in spring that is packed inside a parachute container. When the ripcord is pulled, the container opens and the spring-loaded pilot chute pushes into the air.

square A rectangular-style ram-air parachute is often called square.

S&TA See **safety and training advisor.**

stable A controlled flight accomplished by manipulating the body in freefall.

static line (S/L)   A lanyard of nylon webbing that connects a jumper's rig to the airplane. As the jumper falls away, the static line opens the main container and facilitates the deployment of the main parachute. See also **direct bag** and **pilot chute assist.**

TAF (Tandem Accelerated Freefall)   One of several hybrid training programs developed by individual skydiving schools, often featuring one or more tandem jumps followed by harness-hold jumps.

tandem   A parachute rig built for two people and usually used for training.

tandem instructor   A skydiver who has been certified in the use of a tandem parachute system. Also called a tandem master or parachutist in command.

terminal velocity   The constant speed of a falling person or object. The speed will be dependent on such things as weight, surface area, and body position. In any given position, a freefalling skydiver will accelerate to a specific speed, and then no further acceleration is possible without changing body position.

throw-out pilot chute   A pilot chute that is pulled from a pocket and thrown into the air to begin the opening of a parachute. A throw-out pilot chute is usually contained in a pocket either on a jumper's leg strap, or the bottom of the parachute container. See also **hand-deploy pilot chute.**

TLO (targeted learning objective)   A specific goal for a training skydive.

tracking   Flying horizontally across the sky by retracting the arms and extending the legs.

tunnel rat   A person who spends lots of time in a vertical wind tunnel (slang).

turbine engine   A type of high-performance engine used on some skydiving airplanes.

USBA (United States BASE Association) A very informal group that supports and coordinates some BASE jumping activity.

USPA (United States Parachute Association) A voluntary industry group responsible for providing standards for skydiving in the United States. The USPA certifies instructors and issues licenses. Most skydivers are members of the organization, and about 80 percent of drop zones in the United States have agreed to follow rules, regulations, and standards published by this organization.

Vector A brand of tandem rig manufactured by The Uninsured Relative Workshop.

vRW (vertical relative work) Flying your body in the vertical plane rather than horizontal. vRW generally means in a head-down position.

waiver A legal document that relinquishes certain rights. Skydiving schools almost always require students to sign a waiver that prohibits recovery of damages in the event of an accident.

WDI (wind drift indicator) A device thrown from an airplane prior to jumping to gain an understanding of the wind conditions. A WDI is usually a long piece of crepe paper with a wooden stick, metal rod, or weight on one end.

whuffo A person who does not skydive (slang).

wind dummies The first people to jump each day when a WDI is not used (slang). If the drop zone does not throw a WDI, there will be very little information about the specific wind conditions, and the first few jumpers to fly their parachutes will need to determine the wind drift on their own.

wind tunnel A vertical tunnel used for skydiving training. A powerful fan in the floor or ceiling creates a wind that simulates freefall.

wingsuit   A skydiving suit with extra material between the arms and the body, and between the legs. The material slows the jumper's descent rate and makes it possible to travel greater distances over the ground.

wrap   When two or more parachutes collide or become entangled in the air.

zero porosity   A special kind of coated nylon used for parachutes. Sometimes called ZP.

Acknowledgments

Writing a book is a challenge. This project, however, was greatly simplified by the assistance of many leaders in the skydiving community. I wish to express my gratitude to all of the folks I spoke with during the early stages of the project, including T. K. Donle of the Uninsured Relative Workshop, Jim Crouch and Ed Scott of the United States Parachute Association, Nancy LaRiviere of Jump Shack, Gary Speer of the Flyaway Wind Tunnel, Omar Arias of the SkyVenture wind tunnel, Paul Fayard of Carolina Sky Sports, Ann Maxwell of Skydive University, and Bill Dause of The Parachute Center in California. The staffs at Skydive Wayne County in Richmond, Indiana, and Cleveland Sport Parachute Center provided great insight into their experience with the Sport ParaSim. I am grateful to John Tanis for the chance he gave me to spend time in an outdoor wind tunnel called Extreme, which travels the college festival circuit.

I am fortunate to be a part of the staff at The Ranch Parachute Center in Gardiner, New York, just a bit north of New York City. Teaching offered me a great opportunity to spend time with many students who shared their own first jump stories and anxieties. Their thoughts and comments provided direction, and helped shape this book. I also owe thanks to the staff at The Ranch for their assistance throughout the busy summer season, and for helping me identify issues that are important to first-time jumpers.

I am especially indebted to Kamuran "Sonic" Bayrasli, who graciously offered the loan of his digital camera and provided a few of the photos that are included in the book. "Sonic" owns The PROshop, an on-site equipment dealership at The Ranch, and his shop frequently became the hub of activity as photos were swapped between cameras and computers.

Graphics were produced under tight deadline pressure by Laura Maggio. Photography was provided by a wide range of talented shooters from a variety of skydiving centers. Free-fall photography is an especially demanding specialty, and many of the photographers who provided photos for this project are among the best in the world. I owe them huge thanks, and hope that readers will appreciate the special skill and talent each has brought to the sky.

The manuscript was reviewed at various points by T. K. Donle and Bill Booth of The Uninsured Relative Workshop; Jim Crouch, Ed Scott, and Kevin Gibson of the United States Parachute Association; Mike Truffer of *Skydiving* magazine; and skydiving instructor Mike Lanfor. Some of the chapter material was reviewed by Max Cohn of Generation Freefly, John Ulczyckji of the National Safety Council, and John J. DeRosalia, who is the author of *Mental Training for Skydiving and Life*. Each provided interesting review notes that helped shape the final manuscript.

I must also extend appreciation to George Woods, my first jump instructor at Frontier Skydivers, near Buffalo, New York, as well as the many instructors and skydivers with whom I have been privileged to fly over the last 20-plus years. Their efforts and insights have shaped my teaching style, and truly form the foundation of this book.

Index

Note: Boldface numbers indicate illustrations.

Accelerated Freefall (AFF), 35–38, 46, 48–49, 196
  altitude for, 35–36
  arched body position for, 36, 136
  call-outs during, 79
  communications during, 36, **37**
  freefall simulators for, 39–44
  ground training for, 37, **38**
  instructors for, 59
  lazy w technique in, 136
  skill tests, 68–69
  special maneuvers during, 37–38
  speed or time of fall in, 36, 197–198
  training for, 36–37
Accelerated Freefall rating (USPA), 59
accidents and fatalities, 12, 119–132, 186–187, 199
  causes of, 123–126
  injuries and, 128
  landings and, 125–126
  malfunctions of equipment and, 123–124
  other activities vs. skydiving in, 128–130
  prevention of, 122–123
  tandem jumps and, 126–128, **127**
  USPA reports on, 120–126, **122**
advanced ground school (AGS), 46
age of skydivers, 6–7, **6**, 202
aircraft, 107–118
  capacity of, 190, 200
  CASA 212, 115–116
  Cessna 182, 111–113, **112**

aircraft (*Cont.*):
  Cessna Caravan, 113–114
  common jump planes, 110–116
  helicopters, balloons, and others, 117
  King Air, 114
  Pilatus Porter, 113
  piston versus turbine type, 107–110
  seatbelt requirements in, 109–110
  Skyvan, 115–116
  Twin Otter, 114, **115**
  weight distribution in, 110
airspace restrictions, 52
Airtech GmbH, 97
alcohol and skydiving, 161–162, 196
altimeters and instrumentation, 102–105, **103**, 189–190
altitude, 57–58, 190
  Accelerated Freefall (AFF) and, 35–36
  aircraft types and, 107–110
  altimeters and instrumentation for, 102–105, **103**
  decompression sickness or the bends and, 148
  formation skydiving and, 138
  high altitude jumps and, 147–148
  oxygen masks and, 148
  speed jumps and, 143–144
  static line jumps and, 33–34
  tandem jumps and, 28–30
altitude-recording computers, 104, **104**
anxiety, 17
arched body position, 36, 78–79, 134–136, **134**
Astra AADs, 96–97

automatic activation device (AAD),
    23–24, 54, 57, 84, 94, 96–99,
    123–124, 126, 189
  manufacturers and types of, 96–99
  maintenance of, 98–99

balloons as jump aircraft, 117
BASE jumping, 152–154, **153**
Basic Safety Requirements (BSRs), 55,
    57–58, 94
beer rules, 161–162
biplanes as jump aircraft, 117–118, **118**
blind persons and skydiving, 14–15
boogie parties, 160–161, 167
breathing, 16, 204, 205
bridges, jumping from (See BASE
    jumping)
building, antenna, span, and earth (See
    BASE jumping)
Bush, George, as skydiver, 35, 45

call-outs, 78–79, 201
camping, 160
canopy parachutes, 148–149
canopy relative work (CRW), 150–152
car skydiving, 116, **116**
CASA 212 jump plane, 115–116
causes of accidents/fatalities, 123–126
certification, schools and instructors,
    199
Cessna Caravan jump planes, 113–114
Cessna 182 jump plane, 111–113, **112**
clear and pull technique, static line
    jumps and, 34
cliffs, jumping from (See BASE jumping)
clothing for skydiving, 203
clouds, 9, **10**, **53**, 201–202
Coach rating, 58, 60–61, 173–174
commands, call-outs, 78–79
communications, 201
comprehensive ground school (CGS), 46
controls and handles, 89
costs, 7–8, 179, 191, 198
    aircraft-associated, 108
    drop zone, 165, 166
    equipment, 90, 94, 95
Course Director (USPA), 59

creepers for training, 139, **140**
CRW (See canopy relative work)
cutaway handles, 93–94
Cypres (Cybernetic Parachute Release
    System) AADs, 96–97, **97**

decompression sickness, 148
disabled persons and skydiving, 14–15,
    202
discounts, 198
drogues, 29–30, 87–89, **88**
drop zones (and schools), 4–5, 46,
    57–58, 157–176, 179–194
  accident history of, 186–187
  aircraft provided by, 190
  alcohol (beer rules) and, 161–162
  altitude of jumps at, 190
  checking in to, 164–165
  coaches for, 173–174
  comparison shopping for, 191–194
  competition among, 158–159
  contacting, by phone, 182–183
  costs associated with, 165, 166, 191
  equipment provided by, 189
  history and reputation of, 187
  instructors and, 166–168, 174,
    184–185
  jobs required for, 168–176
  large-scale, 162–165
  licensing of, 199, 206
  locating, 177–181, 195
  manifest for, 169
  number of, in U.S., 207
  number of students trained at, 187
  organizing and running a, 157–159
  packer for, 169–171, **170**
  photographers for, 172–173, **173**
  pilots for, 175
  questions to ask before jumping,
    182–192
  riggers for, 170, 171–172
  safety equipment and, 189
  seasonal operations in, 158, 190–191
  selecting, 177–178
  Skydive University programs at,
    188–189
  small-size, 165–168

drop zones (and schools) (*Cont.*):
  social activities for, 160–162
  spectators and, 159–160, **167**
  state regulations concerning, 61–62
  training methods offered by, 188
  USPA group membership, 183–184
  USPA instructor ratings, 184, 193
  Web sites for, 180, 181
Dual Hawk tandem rig, 90

ear congestion, 17–18, 196
eating before a jump, 204
Eclipse tandem rig, 90
electronic AADs, 96–97
emergencies, 23–24, 198–199
empowerment through skydiving, 80–81
environment and fear management,
      73–74
equipment, 72–73, 83–106
  age of, 189
  altimeters and instrumentation,
      102–105, **103**, 189–190
  automatic activation device (AAD),
      84, 94, 96–99, 189
  clothing and, 203
  complete rig, 83–84, **85**
  controls and handles, 89
  costs of, 90, 94, 95
  cutaway handles, 93–94
  drogues, 87–89, **88**
  expert rigs, 94–95
  goggles, 101–102
  harness, 84, 86–87, **86, 87**
  helmets, 100–101, **101**
  jumpsuits, 99
  log books, 105–106, **105**
  malfunctions of, in accidents/fatalities,
      123–124
  manufacturers of, tandem-type, 90
  parachute materials, 199
  pilot chutes, 91, **92**
  reserve chute, 84
  reserve static line (RSL), 89, 94, 189
  ripcords, 89
  second-hand, 95
  single operation system (SOS) rigs,
      94

equipment (*Cont.*):
  square (rectangular, ram air)
      parachutes, 91–93
  student parachute rigs, 90–94
  suppliers of, 84
  tandem-type, 62–64, 84–90
  weight limits for, 12–14, 202–203
  weight of, 203
Evaluator (USPA), 59
excitement, 79–80
exhibition jumps, 152
experience level vs. risk, 122, **123**
expert skydiver rigs, 94–95
eye protection, 101–102
eyeglasses, 202

fainting, 23, 197
fatalities (*See* accidents and fatalities)
fear and fear management, 1, 65–76,
      197
  call-outs to help in, 78–79
  converting stress to excitement in,
      79–80
  empowerment through, 80–81
  environment and, 73–74
  equipment and, 72–73
  friends' help with, 75–76
  others in model of, 69–72
  relaxation techniques for, 76–78, **77**
  self in model of, 67–69
fear of heights, 66, 197
Federal Aviation Administration (FAA)
      regulations, 12–13, 25, 51–54,
      62–64, 127–128
Flyaway Indoor Skydiving, 41–42
flying the parachute and, 137
formation skydiving, 137–140, 149–152
  canopy relative work (CRW) and,
      149–152
  wraps or entanglement problems in,
      151–152
four-way freefall, 138, **138**
freefall simulators, 39–44
freefly technique in, 140–142, **141**
frequently asked questions (FAQs),
      195–204
FXC 12000 AAD, 97–98

gender of skydivers, 4, **4**
glory effect, 9
goggles, 101–102, 202
grippers, for formation skydiving, 139
ground training, 37, **38**, 46
groups for skydiving, 5–7

hand signals, 36, **37**, 201
hard arch, 135, **135**
harness, 84, 86–87, **86**, **87**
hearing-impaired and skydiving,
    14–15
heights, fear of, 66, 197
helicopters as jump aircraft, 117
helmets, 57, 100–101, **101**
high altitude jumps, 147–148
human flight, 133–155

indoor skydiving, 41–42
injuries (*See* accidents/fatalities)
Injury Facts, 128–129
instructor assisted deployment (IAD),
    32, 59
Instructor rating (USPA), 58–59
Instructor-Examiner rating (USPA),
    59
instructors, 69–72, 166–168,
        184–185
    income of, 174
    selecting, 200
    tipping, 198
insurance, 11–12, 55, 203
Integrated Student Program (ISP), 58

Jump Shack, 90
jumpsuits, 8–9, 99, 142
    formation skydiving and, grippers on,
        139
    wingsuits and, 144–145, **145**

King Air jump planes, 114

landings, 89, 125–126, 198
lazy w technique, 136
liability and insurance, 11–12, 203
licenses, 56–57, 199, 206
log books, 105–106, **105**

maintenance of AADs, 98–99
making your decision, 177–194
manifest, for drop zone, 169
marriage proposals, 5–6
mechanical AADs, 97–98
medical and health issues, 14–18,
        196–197, 202, 205
motion sickness, 17

National Safety Council (NSC), 128–129
nausea, 17, 196–197
New River Gorge Bridge, WV, 154
numbers of skydivers in U.S., 3, 121

obstructions to drop zone, 57–58
opening of parachute, speed of fall and,
        21–22, 199, 200–201
Otter jump planes, 114, **115**
oxygen, oxygen masks, 148, 205

parachute packing and packers,
        regulation of, 53–54, 169–171,
        170, 186, 201
parachute simulators, 44–46
parachute swooping, 148–149
parachutes, 199
    (*See also* equipment)
permits, state issued, 61–62
personality of skydivers, 3–5
photographers and photography,
        20–21, 139, 159–160, 172–173,
        173, 203
Pieces of Eight, 15
Pilatus Porter jump planes, 113
pilot chutes, 32, 91, **92**
pilots, 175, 186
pressure changes, ears and sinus and,
        17–18
preventing accidents, 122–123
psychology, 65–81

questions to ask, evaluating a drop
        zone/school, 182–192

ram air parachute, 91–93
ratings, USPA, 58–60, 184, 186, 199,
        206

reasons to skydive, 3
rectangular parachute, 91–93
regulation of skydiving, 51–54
    (*See also* Federal Aviation
        Administration)
relaxation techniques, 76–78, **77**
reserve parachutes, 23–24, 84, 198–199,
    201
reserve static line (RSL), 57, 89, 94,
    126, 189
rig, complete, 83–84, **85**
rig, student parachute type, 90–94
riggers, 53–54, 170, 171–172, 186, 201
ripcords, 89, 200–201
risk, 66–67, 119–132, 186–187, 199
    (*See also* accidents/fatalities)

safety, 55, 57–60
Safety and Training Advisor (S&TA),
    59–60, **59**
schools, 4–5, 179–194
    (*See also* drop zones)
seatbelts in aircraft, 109–110
second-hand equipment, 95
self-confidence in skydiving, 81
self in fear management, 67–69
sensory overload, 18–20
shape of parachute, 22–23, 91–93, 199
Sigma tandem rig, 90
single operation system (SOS) rigs, 94
sinus blockage, 17–18, 196
size of parachutes, 94–95
sky surfing, 146–147, **146**
Skydive University, 60–61, 188–189, 200
Skydivers Information Manual (SIM),
    56
Skydivers Over Sixty (SOS), 7
*Skydiving for Wheelchair Dependent
    Persons*, 15–16
Skyvan jump planes, 115–116
SkyVenture wind tunnels, 42–44, **43**
social activities at drop zones,
    160–162
spectators, 159–160, **167**, **196**
Specter of the Brocken, 9
speed, 9–10, 16, 20–23, 142–144,
    197–198

speed (*Cont.*):
    Accelerated Freefall (AFF) and, 36
    drogues and, 29–30, 88
    freeflying and, 141–142
    opening/at opening of parachute,
        21–22, 201
    shape of parachutes and, 22–23
    size of parachute and, 94–95
    speed jumps, 142–144, **143**
    static line jumps and, 34–35
    swooping and, 149
    tandem jumps and, 28–30
speed skydiving, 142–144
square parachutes (rectangular, ram air),
    91–93
stack formation, 150–151, **151**
state regulations, 61–62
static line jumps, 30–35, **31**, 195
    altitude for, 33–34
    clear and pull technique in, 34
    instructor assisted deployment (IAD)
        in, 32
    instructors for, 59
    pilot chute used in, 32
    radioed instructions in, 33, **33**
    skill tests before, 68–69
    speed of, 34–35
    training for, 33–35
statistics on skydivers, 3–5, **4**, 6–7, **6**, 121
steering toggles, 22, 92–93, **93**
stress (*See* fear and fear management)
Strong Enterprises, 90
student parachute rigs, 90–94
Stunts Adventure Equipment, 90
surfing, sky, 146–147, **146**
swooping, 148–149

Tandem Instructor rating (USPA), 59
tandem jumps, 1–2, 25–30, **27**, 47, 48,
    55, 69–70, 195, 196
    accidents/fatalities and, 126–128, **127**
    altitude and speed in, 28–30
    controls and handles in, 89
    costs of equipment for, 90
    drogues in, 87–89, **88**
    equipment for, 13–14, 62–64, 84–90,
        84, **85**

tandem jumps (*Cont.*):
 FAA regulations and, 127–128
 harness for, 86–87, **86**
 instructors for, 59
 landings and, 89
 manufacturers of equipment for, 90
 regulations concerning, 54
 reserve static line (RSL) and, 89
 ripcords in, 89
 speed and, 88
 training required for, 26–28
temperatures, 148
testing your skills, 68–69
tipping instructors **174**, 199
toggles for steering, 22, 92–93, **93**
tracking technique, 49
training methods, 25–50, 164–165, 179, 188
 for Accelerated Freefall (AFF), 35–38, 46, 48–49
 advanced ground school (AGS) as, 46
 advanced techniques and skills in, 49–50
 comprehensive ground school (CGS) as, 46
 creepers in, 139, **140**
 Flyaway Indoor Skydiving as, 41–42
 freefall simulators for, 39–44
 ground training as, 46
 hybrid programs for, 45–50
 instructor assisted deployment (IAD) in, 32
 Integrated Student Program (ISP) in, 58
 parachute simulators for, 44–46
 progressing with different methods in, 38–39
 skills developed during, 47–48
 Skydive University and, 60–61

training methods (*Cont.*):
 SkyVenture wind tunnels for, 42–44, **43**
 for static line jumps, 30–35, **31**
 for tandem jumps as, 25–30, **27**, 47, 48
 time required for, 198
 video-, 39
 wind tunnels and simulators in, 40–44, **41**
turbulence, 10–11
turning in flight, 136
Twin Otter jump planes, 114, **115**
two-way freefall, 137–138

Uninsured Relative Workshop, The, 15, 90
United States Parachute Association (USPA), 54–60
 accident/fatality reports by, 120–126, **122**
 drop zone location listing, 179, **180**
 drop zone membership in, 183–184, 193
 ratings by, 58–60, 184, 186, 193, 199, 206

Vector tandem rig, 90
video training programs, 39
videotaping a jump, 20–21, **21**, 159–160
virtual reality skydiving, 44–46, **45**
vision problems, 202

weather, 8–11, 202
Web sites of interest, 180, 181
weight limits, 12–14, 202–203
wind, 10–11, 57–58, 202
wind tunnels and simulators, 40–44, **41**
wingsuit flight, 144–145, **145**
wraps or entanglements, 151–152